Springer Tracts on Transportation and Traffic

2

Series Editor

Roger P. Roess
Polytechnic Institute of New York University
Metrotech Center 6
11201 New York
USA
E-mail: rroess@poly.edu

For further volumes:
http://www.springer.com/series/11059

Konstantinos Tzanakakis

The Railway Track and Its Long Term Behaviour

A Handbook for a Railway Track of High Quality

Springer

Author
Konstantinos Tzanakakis
Athens
Greece

ISSN 2194-8119 e-ISSN 2194-8127
ISBN 978-3-642-36050-3 e-ISBN 978-3-642-36051-0
DOI 10.1007/978-3-642-36051-0
Springer Heidelberg New York Dordrecht London

Library of Congress Control Number: 2012955945

Printed on acid-free paper

Springer is part of Springer Science+Business Media (www.springer.com)

"It appears incontrovertible that understanding failure plays a key role in error-free design of all kinds, and that indeed all successful design is the proper and complete anticipation of what can go wrong."

Henry Petroski
Design Paradigms
Case Histories of Error and Judgment in Engineering

"Quality is never an accident; it is always the result of high intention, sincere effort, intelligent direction and skilful execution; it represents the wise choice of many alternatives"

William A. Foster

"You have to learn the rules of the game. And then you have to play better than anyone else"

Albert Einstein

Foreword

Trains have to run on time, and a basic condition is that the quality of the track and other infrastructure objects is sufficient. Most of the physical systems and their components degrade/fail over time. This affects the performance of the system and can lead to system failures. Failure during operation can be costly, e.g. loss of service, property and even life.

The track engineer when he constructs, renews or maintains a track is of course aiming to deliver a track of high quality: but how does the quality affect the track costs and the service life of the track?

There is a lack of understanding of the relationship between track design, construction, usage and maintenance costs. Therefore it is of greatest importance to create common and better understanding of these relations.

Albert Einstein said "***you have to learn the rules of the game. And then you have to play better than anyone else***". The book aims to provide the necessary functional knowledge of the track behaviour, covering the function of the various track components, their interaction as elements of the track system, and the interaction of the systems track and railway vehicles. When we understand how the whole system functions (railway vehicles running on a track), taking into account its imperfections and showing how they influence the quality and performance of the system, then we learn the rules of the game, and we can "play better" in order to improve the efficiency of the Infrastructure Manager (IM), to improve the quality of the track for an optimum long term behaviour, using the available tools (i.e. RAMS, LCC) and choosing the best maintenance strategies (incl. any appropriate outsourcing strategies).

The aim of the IM is to design, construct and maintain an expensive asset, as it is the railway track, in a cost effective way, to assure its high performance during its service life. The only way to achieve this target is included in the word "**Quality**". If the Infrastructure Manager can achieve to get a new track of high quality, to maintain it with the optimum maintenance strategies, he will be able to manage a track with low costs and a high efficiency.

The IM has to act in an external environment, including the stakeholders, mainly the government and the society, which set increasing demands on the IM by providing at the same time less funding, as often foreseen in multiannual contracts between the State and the IM. The splitting of the traditional (integrated) railways in management of the infrastructure (responsibility of the IM) and operation (responsibility of the Railway Undertakings or Operators) introduced a new "player", the Railway Undertakings, who also set -through the contractual agreements with the IM- high demands on track efficiency.

Hence, the IM is under considerable pressure to act in a high demanding external environment, with a much lower budget than in the past. The tools for the

IM to manage a qualitative, efficient track with good long term behaviour are also being here presented here.

The book is structured in five parts (A, B, C, D, E) and 10 sections, as presented below. Confucius said "study the past if you would define the future". So, it starts with the Past (railway history), then the external environment for the IM is given describing the context in which the IM is acting, followed by the functional knowledge to understand how and why does the track system works as it works, tools and methods aiming at high performance, and finally it ends with the Future (the future challenges for the IM), consistent with the quote of Confucius. The following figure gives a respective overview.

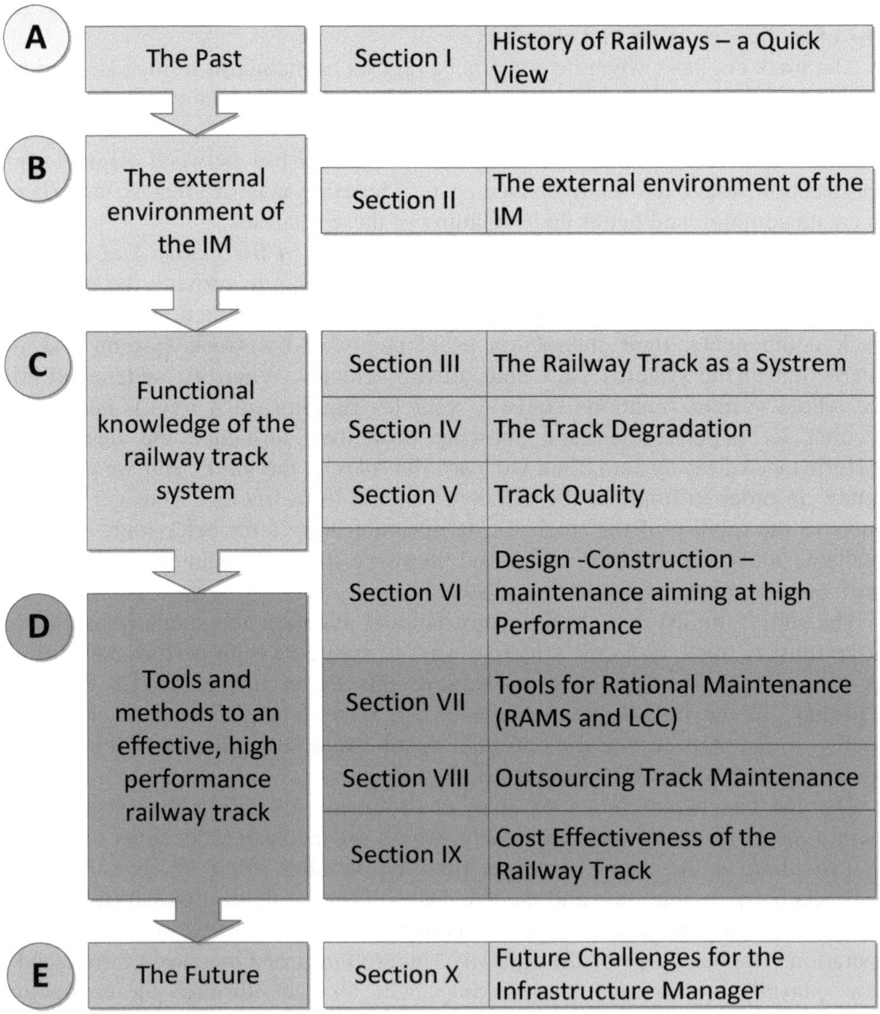

Athens, December 2012 Konstantinos Tzanakakis

Overview

Part A: The Past	
Section I	**History of Railways – A Quick View** This section gives a quick view of railway history: The history of Railways dates back nearly 500 years, and includes systems with man or horse power and rails of wood or stone. Modern rail transport systems first appeared in England in the 1820s. These systems, which made use of the steam locomotive, were the first practical forms of mechanized land transport, and they remained the primary form of mechanized land transport for the next 100 years.
Part B: The external environment of the IM	
Section II	**The External Environment of the Infrastructure Manager (IM)** The Infrastructure Manager has to act in an increasingly challenging external environment. There are various stakeholders who have interests in the activities of the railway Infrastructure Manager. The available government budgets are decreasing, and at the same time there are increased needs for higher infrastructure capacity utilisation. An efficient way of managing the rail infrastructure is required.
Part C: Functional Knowledge of the Railway Track System	
Section III	**The Railway Track as a System** In this section the knowledge about the "how" and "why" of the track behaviour is provided, as a base knowledge for highlighting the importance of track quality.

Section IV	**The Track Degradation** In this section the track deterioration mechanisms are discussed, which lead to a degradation of the track, and also the factors influencing the durability of the track geometry. Recommendations and good practices for preventing the track degradation are also given.
Section V	**Track Quality** Quality should be the final goal for the Infrastructure Manager. In this Section we examine the importance of good quality and how quality affects the long term behaviour of the railway track.
Section VI	**Design -Construction – Maintenance Aiming at High Performance** A high performance railway track has to be designed and constructed properly. Consequently, when trains are running on the track (operational phase), the track has to be monitored and maintained. What are the possible maintenance strategies? Which strategy does best fits the needs and the culture of the various Infrastructure managers?
Part D: Tools and Methods for an Effective, High Performance Railway Track	
Section VII	**Tools for Rational Maintenance (RAMS and LCC)** In this section tools for the optimization of the track construction regarding technical and economic requirements are presented. LCC and RAMS technology are two acknowledged methods for assisting the optimization process regarding those requirements.
Section VIII	**Outsourcing Track Maintenance** Outsourcing of maintenance services is not a definite path to maintenance excellence. Through outsourcing it could be possible to cut cost and make the maintenance more effective. In this section, advantages and disadvantages of outsourcing, factors influencing decision making, types and characteristics of outsourcing contracts, and cases of applied outsourcing are topics discussed among others.

Section IX	**Cost Effectiveness of the Railway Track** The infrastructure as the main production factor in the railway system's value chain accounts for a significant part of the full system costs. Therefore it is very important for Infrastructure Managers to have a good control of infrastructure management in general and of costs in particular to improve competitiveness. Aspects that are discussed in this section include guiding principles for an effective railway track, methods to measure track performance, benchmarking as a practical tool for improving performance by learning from best practices and understanding the procedures by which they are achieved. Furthermore, a method for estimating costs related to maintenance and renewal of the track are also addressed.
Part E: The Future	
Section X	**Future Challenges for the IM** Today more trains are running on the track than in the past and the competition with other means of transportation is becoming greater. At the same time there is also a tendency towards decreased time and funds for maintenance. The characteristics of the future railway as a system will be discussed in this section.

Contents

Part C: Functional Knowledge of the Railway Track System

List of Figures

List of Tables

Part A
The Past

"Study the past if you would define the future"
Confucius

"What you need to know about the past is that no matter what has happened, it has all worked together to bring you to this very moment. And this is the moment you can choose to make everything new. Right now."

"You have to know the past to understand the present."

Dr. Carl Sagan

Section I: History of Railways – A Quick View

This section gives a quick view of railway history: The history of railways dates back nearly 500 years, and includes systems with man or horse power and rails of wood or stone. Modern rail transport systems first appeared in England in the 1820s. These systems, which made use of the steam locomotive, were the first practical forms of mechanized land transport, and they remained the primary form of mechanized land transport for the next 100 years.

1 Earliest Traces

The idea of using "tracked" roads is at least 2000 years old. Ancient civilisations realised that wheeled vehicles ran more efficiently and needed less maintenance if they were guided using grooves cut into the stone blocks of roads. Quarries in Ancient Greece, Malta and the Roman Empire used cut stone tracks to haul loads pulled by animals.

Evidence of these first 'rail lines' used to guide chariots and wagons of the Greek and Roman Empires can be seen today at ancient sites in Greece and Italy.

The earliest evidence of a wagon way, a predecessor of the railway, found so far was the 6 to 8,5 km long Diolkos wagon way, which transported boats across the Isthmus of Corinth in Greece since around 600 BC (Figure 1 and Figure 2).

Fig. 1 Diolkos

Wheeled vehicles pulled by men and animals ran in grooves in limestone, which provided the track element, preventing the wagons from leaving the intended route. The Diolkos was in use for over 650 years, until at least the 1st century AD. The first horse-drawn wagon ways also appeared in ancient Greece, with others to be found in Malta and various parts of the Roman Empire, using cut-stone tracks.

K. Tzanakakis: *The Railway Track and Its Long Term Behaviour*, STTT 2, pp. 5–7.
DOI: 10.1007/978-3-642-36051-0_1

Fig. 2 Diolkos: transport of boats across the Isthmus of Corinth in Greece

Fig. 3 Reisszug, a funicular railway at the Hohensalzburg Castle in Austria

Railways began reappearing in Europe after the Dark Ages. The earliest known record of a railway in Europe from this period is a stained-glass window in the Minster of Freiburg im Breisgau dating from around 1350.

In 1515, Cardinal Matthäus Lang wrote a description of the Reisszug, a funicular railway at the Hohensalzburg Castle in Austria (Figure 3). It is notable for its extreme age, as it is believed to date back to either 1495 or 1504.The line originally used wooden rails and a hemp haulage rope, and was operated by human or animal power, through a tread wheel. The line still exists, albeit in updated form, and is probably the oldest railway still to operate.

2 Early Wagonways

Fig. 4 Section of timber track from a 16th-century gold mine in Siebenbürgen. The wagons were guided by the pronounced flange on the wooden wheels, and the narrow gauge of 48 cm allowed the points to be altered by swinging the single switch rail.

Wagon ways (or 'tramways') are thought to have developed in Germany in the 1550s to facilitate the transport of ore tubs to and from mines, utilising primitive wooden rails (Figure 4). Such an operation was illustrated in 1556 by Georgius Agricola. These used the “hund” system with unflanged wheels running on wooden planks and a vertical pin on the truck fitting into the gap between the planks, to keep it going the right way. Such a transport system was used by German Miners at Caldbeck, Cumbria, perhaps from the 1560s.

The first true railway is now suggested to have been a funicular railway made at Broseley in Shropshire at some time before 1605. This carried coal for James Clifford from his mines down to the river Severn to be loaded on to barges and carried to riverside towns. Though the first documented record of this is later, its construction probably preceded the Wollaton Wagon way, completed in 1604, hitherto regarded as the earliest British installation. This ran from Strelley to Wollaton near Nottingham. Another early wagon way is noted onwards. Huntingdon Beaumont (who was concerned with mining at Strelley) also laid down broad wooden rails near Newcastle upon Tyne, on which a single horse could haul fifty or sixty bushels (130–150 kg) of coal.

By the eighteenth century, such wagon ways and tramways existed in a number of areas. Ralph Allen, for example, constructed a tramway to transport stone from a local quarry to supply the needs of the builders of the Georgian terraces of Bath. The Battle of Prestonpans, in the Jacobite Rebellion, was fought astride a wagon way. This type of transport spread rapidly through the whole Tyneside coal-field, and the greatest number of lines were to be found in the coalfield near Newcastle

K. Tzanakakis: *The Railway Track and Its Long Term Behaviour*, STTT 2, pp. 9–10.
DOI: 10.1007/978-3-642-36051-0_2

upon Tyne, where they were known locally as wagon ways. Their function in most cases was to facilitate the transport of coal in chaldron wagons from the coal pits to a staithe (a wooden pier) on the river bank, whence coal could be shipped to London by collier brigs. The wagon ways were engineered so that trains of coal wagons could descend to the staithe by gravity, being braked by a brakeman who would "sprag" the wheels by jamming them. Wagon ways on less steep gradients could be retarded by allowing the wheels to bind on curves. As the work became more wearing on the horses, a vehicle known as a dandy wagon was introduced, in which the horse could rest on downhill stretches.

Fig. 5 Landscape photograph (taken during excavations) looking along the length of the Lambton Waggon way showing sleepers, rails and points (Image from 9 March 1996) (University of Newcastle)

3 Horse-Drawn Railway

In medieval times people mostly travelled on foot or horseback and any form of transportation was mainly for moving goods.

The first railways were laid down in the seventeenth and eighteenth century for horse drawn trains of wagons in collieries and quarries. These 'hauling ways' initially had a surface of stone slabs or timber baulks, which soon proved unsatisfactory as the loads carried inevitably grew heavier.

As the Industrial Revolution progressed, the idea was developed further by adding cast iron or wrought iron plates to reduce wear on the wooden baulks. This evolved further to iron edge rails enabling the use of flanged wheels for the first time.

Placing a horse car on rails had enabled a horse to move twice as many people, and so street railways were born.

Fig. 6 Horse-drawn railway, designed for personal transport

K. Tzanakakis: *The Railway Track and Its Long Term Behaviour*, STTT 2, pp. 11–12.
DOI: 10.1007/978-3-642-36051-0_3

Fig. 7 Horse-drawn railway, ride from Linz to České Budějovice, personal transport I. and II. class, historical graphic work (http://www.ckrumlov.info)

4 Steam Power Introduced

James Watt, a Scottish inventor and mechanical engineer, was responsible for improvements to the steam engine of Thomas Newcomen, before used to pump water out of mines. Watt developed a reciprocating engine, capable of powering a wheel. Although the Watt engine powered cotton mills and a variety of machinery, it was a large stationary engine.

The first working model of a steam rail locomotive was designed and constructed by John Fitch in the United States in 1794.

The first full scale working railway steam locomotive was built in the United Kingdom in 1804 by Richard Trevithick, an English engineer born in Cornwall (the story goes that it was constructed to satisfy a bet by Samuel Homfray, the local iron master). This used high pressure steam to drive the engine by one power stroke (the transmission system employed a large fly-wheel to even out the action of the piston rod) (Figure 8).

On February 21, 1804 Trevithick's engine hauled 10 tons of iron and 70 men nearly ten miles from Penydarren, South Wales at a speed of five miles-per-hour. Trevithick later demonstrated a locomotive operating upon a piece of circular rail

Fig. 8 A replica of Trevithick's engine at the National Waterfront Museum, Swansea

K. Tzanakakis: *The Railway Track and Its Long Term Behaviour*, STTT 2, pp. 13–16.
DOI: 10.1007/978-3-642-36051-0_4

track in Bloomsbury, London, the "Catch-Me-Who-Can", but never got beyond the experimental stage with railway locomotives, not least because his engines were too heavy for the cast-iron plateway track then in use. Despite his inventive talents, Richard Trevithick died in poverty, with his achievement being largely unrecognized.

The impact of the Napoleonic Wars resulted in (amongst other things) a dramatic rise in the price of fodder. This was the imperative that made the locomotive an economic proposition, if it could be perfected.

The first commercially successful steam locomotive was Matthew Murray's rack locomotive Salamanca built for the narrow gauge Middleton Railway in 1812 (Figure 9). This twin cylinder locomotive was not heavy enough to break the edge-rails track, and solved the problem of adhesion by a cog-wheel utilising teeth cast on the side of one of the rails. It was the first rack railway.

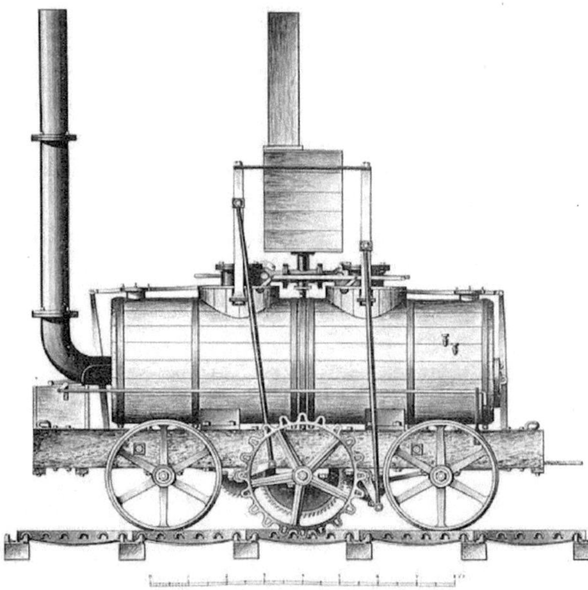

Fig. 9 Salamanca, the first commercially successful steam locomotive, built in 1812 by Matthew Murray of Holbeck

This was followed in 1813 by the Puffing Billy built by Christopher Blackett and William Hedley for the Wylam Colliery Railway, the first successful locomotive running by adhesion only. This was accomplished by the distribution of weight by a number of wheels. Puffing Billy is now on display in the Science Museum in London, the oldest locomotive in existence (Figure 10).

Fig. 10 Puffing Billy as seen from the front (1813-1814)

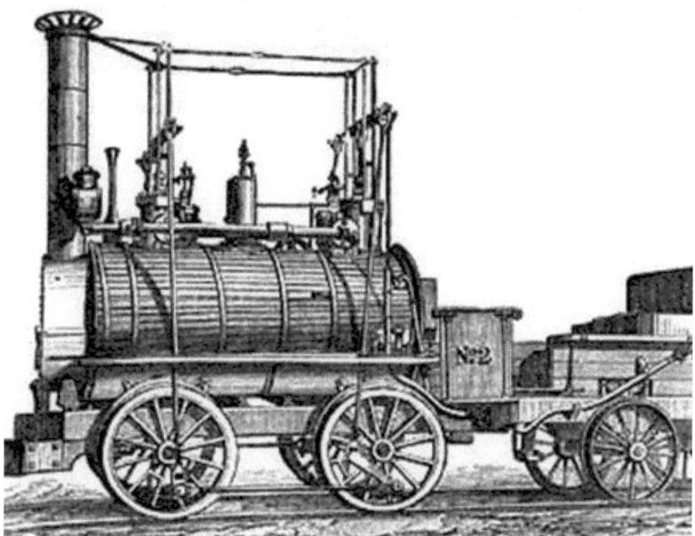

Fig. 11 Blücher engine

In 1814 George Stephenson, inspired by the early locomotives of Trevithick, Murray and Hedley, persuaded the manager of the Killingworth colliery where he worked to allow him to build a steam-powered machine. He built the Blücher, one of the first successful flanged-wheel adhesion locomotives (Figure 11).

Stephenson played a pivotal role in the development and widespread adoption of the steam locomotive. His designs considerably improved on the work of the earlier pioneers. In 1825 he built the Locomotion for the Stockton and Darlington Railway, north east England, which was the first public steam railway in the world (Figure 12). Such success led to Stephenson establishing his company as the pre-eminent builder of steam locomotives used on railways in the United Kingdom, United States and much of Europe.

Fig. 12 Locomotion (at Darlington Railway Centre and Museum)

5 Diesel and Electric Engines

5.1 Electric Railways Revolutionize Urban Transport

In January 1888, Richmond, Virginia served as a proving ground for electric railways as Frank Sprague built the first working electric streetcar system there. By the 1890s, electric power became practical and more widespread, allowing extensive underground railways. Large cities such as London, New York, and Paris built subway systems. When electric propulsion became practical, most street railways were electrified. These then became known as "streetcars," "trolleys," "trams" and "Straßenbahn." They can be found around the world.

In many countries, these electric street railways grew beyond the metropolitan areas to connect with other urban centres. In the USA, "electric interurban" railroad networks connected most urban areas in the states of Illinois, Indiana, Ohio, Pennsylvania and New York. In Southern California, the Pacific Electric Railway connected most cities in Los Angeles and Orange Counties, and the Inland Empire. There were similar systems in Europe. One of the more notable rail systems connected every town and city in Belgium. One of the more notable tramway systems in Asia is the Hong Kong Tramways, which started operation in 1904 and run exclusively on double-decker trams.

The remnants of these systems still exist, and in many places they have been modernized to become part of the urban "rapid transit" system in their respective areas. In the past thirty years increasing numbers of cities have restored electric rail service by building "light rail" systems to replace the tram system they removed during the mid-20th century.

5.2 Diesel Power

Diesel-electric locomotives could be described as electric locomotives with an on-board generator powered by a diesel engine. The first diesel locomotives were low-powered machines, diesel-mechanical types used in switching yards. Diesel and electric locomotives are cleaner, more efficient, and require less maintenance than steam locomotives. They also required less specialized skills in operation. After working through technical difficulties in the early 1900s, diesel locomotives became mainstream after World War II. By the 1970s, diesel and electric power had replaced steam power on most of the world's railroads.

K. Tzanakakis: *The Railway Track and Its Long Term Behaviour*, STTT 2, p. 17.
DOI: 10.1007/978-3-642-36051-0_5

6 High-Speed Railway

Starting with the opening of the first Shinkansen line between Tokyo and Osaka in 1964, high-speed rail transport, functioning at speeds up and above 300 km/h, has been built in Spain, France, Germany, Italy, the People's Republic of China, Taiwan, the United States, the United Kingdom, South Korea, Scandinavia, Belgium and the Netherlands. The construction of many of these lines has resulted in the dramatic decline of short haul flights and automotive traffic between connected cities, such as the Boston-New York City-Washington, D.C. corridor, London-Paris-Brussels, Madrid-Barcelona, as well as many other major lines. Additionally, with the on-going threat of global warming and energy shortages, high-speed rail is supposed to hold the key to the future of transportation in many of the world's developed countries.

Fig. 13. TGV 4402 (operation V150) reaching 574 km/h

K. Tzanakakis: *The Railway Track and Its Long Term Behaviour*, STTT 2, p. 19.
DOI: 10.1007/978-3-642-36051-0_6

7 History of the Track Gauge

Historically, the choice of gauge was partly arbitrary and partly a response to local conditions. Narrow-gauge railways are cheaper to build and can negotiate sharper curves but broad-gauge railways give greater stability and permit higher speeds.

Sometimes railway companies chose their own gauge, such as the Great Western Railway choosing 2,140 mm (7 ft 0 $\frac{1}{4}$ in).

Early origins of the standard gauge

There is an urban legend that Julius Caesar specified a legal width for chariots at the width of standard gauge, causing road ruts at that width, so all later wagons had to have the same width or else risk having one set of wheels suddenly fall into one deep rut but not the other.

In fact, the origins of the standard gauge considerably pre-date the Roman Empire, and may even pre-date the invention of the wheel. The width of prehistoric vehicles was determined by a number of interacting factors which gave rise to a fairly standard vehicle width of a little under 2 m (6.6 ft). These factors have changed little over the millennia, and are still reflected in today's motor vehicles. Road rutting was common in early roads, even with stone pavements. The initial impetus for the ruts probably came from the grooves made by sleds and slide cars dragged over the surfaces of ancient trackways. Since early carts had no steering and no brakes, negotiating hills and curves was dangerous, and cutting ruts into the stone helped them negotiate the hazardous parts of the roads.

Neolithic wheeled carts found in Europe had gauges varying from 1.30 to 1.75 m (4 ft 3 in to 5 ft 9 in). By the Bronze Age, wheel gauges appeared to have stabilized between 1.40 to 1.45 m (4 ft 7 in to 4 ft 9 in) which was attributed to a tradition in ancient technology which was perpetuated throughout European history. The ancient Assyrians, Babylonians, Persians and Greeks constructed roads with artificial wheelruts cut in rock spaced the wheelspan of an ordinary carriage. Such ancient stone rutways connected major cities with sacred sites, such as Athens to Eleusis, Sparta to Ayklia, or Elis to Olympia. The gauge of these stone grooves was 1.38 to 1.44 m (4 ft 6 in to 4 ft 9 in). The largest number of preserved stone trackways, over 150, are found on Malta.

Some of these ancient stone rutways were very ambitious. Around 600 BC the citizens of ancient Corinth constructed the Diolkos, which some consider the world's first railway (see Chapter 1), a hard *poros* limestone road with grooved tracks along which large wooden flatbed cars carrying ships and their cargo were pulled by slaves or draft animals. The grooves were at 1.67 m (5 ft 6 in) centres.

The Roman Empire actually made less use of stone trackways than the prior Greek civilization because the Roman roads were much better than those of previous civilizations. However, there is evidence that the Romans used a more or

K. Tzanakakis: *The Railway Track and Its Long Term Behaviour*, STTT 2, pp. 21–22.
DOI: 10.1007/978-3-642-36051-0_7

less consistent wheel gauge adopted from the Greeks throughout Europe, and brought it to England with the Roman conquest of Britain in AD 43. After the Roman departure from Britain, this more-or-less standard gauge continued in use, so the wheel gauge of animal drawn vehicles in 19th century Britain was 1.4 to 1.5 m (4 ft 7 in to 4 ft 10 in). In 1814 George Stephenson copied the gauge of British coal wagons in his area (about 1.42 m or 4 ft 8 in) for his new locomotive, and for technical reasons widened it slightly to achieve the modern railway standard gauge of 1.435 m (4 ft 8½ in).

Part B

The External Environment of the Infrastructure Manager

Section II: The External Environment of the Infrastructure Manager

The Infrastructure Manager has to act in an increasingly challenging external environment. There are various stakeholders who have interests in the activities of the railway Infrastructure Manager. The available government budgets are decreasing, and at the same time there are increased needs for higher infrastructure capacity utilisation. An efficient way of managing the rail infrastructure is required.

8 The "Stakeholders" of the Railway System

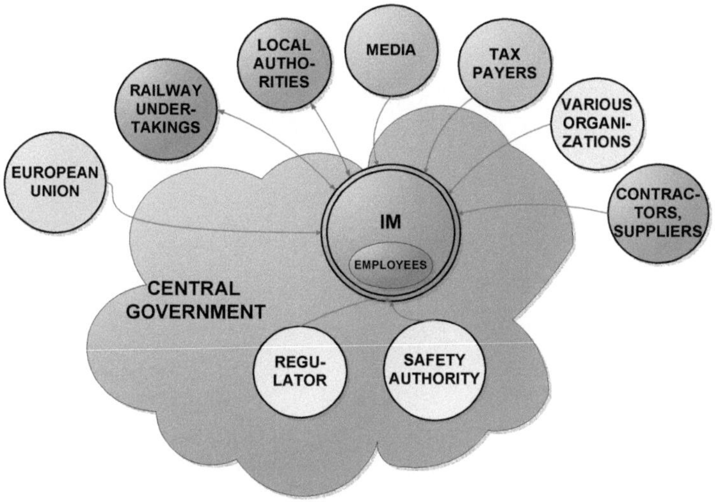

Fig. 14 The stakeholders and the Infrastructure manager (IM)

Table 1 Stakeholders' Objectives (after [22])

Possible Objectives or Priorities	Stakeholders
Reduced Budget Outlays and Lower Subsidies	• Central Government • Local Authorities • Taxpayers
Operational and Cost Efficiency	• Central Government • Local Authorities • Freight Users • Passengers • Private Shareholders
Better Resource Allocation and External Efficiency	• Central Government • Local Authorities • Taxpayers • other Transport Users
Effective Environmental Protection External Efficiency and Optimal Modal-Mix	• Central Government • Local Authorities • Environmentalists
Risk Minimization	• Central Government • Local Authorities • Freight Users • Passengers • Private Shareholders

K. Tzanakakis: *The Railway Track and Its Long Term Behaviour*, STTT 2, pp. 27–28.
DOI: 10.1007/978-3-642-36051-0_8

There are various stakeholders who have interests in the railway Infrastructure Manager (IM) (Figure 14). Table 1 identifies the types of objectives for the different stakeholders.

9 The Changing External Environment for the Infrastructure Manager

In Europe the environment for the rail Infrastructure Managers, is changing:

- new European directives (in the European Union influence zone),
- decreasing government budgets,
- need for higher infrastructure capacity utilisation and
- increasing customer expectations.

This change requires a very efficient way of managing the rail infrastructure maintenance activities.

The current requirements of the external environment to the Infrastructure Manager (IM) can be briefly described as:

⇒ Reduced financing of the IM leads to
  ⇒ Pressure to cost reduction
    ⇒ The IM has to review current practices, find ways to keep and even enhance his performance level by setting new strategies mainly for the track maintenance.

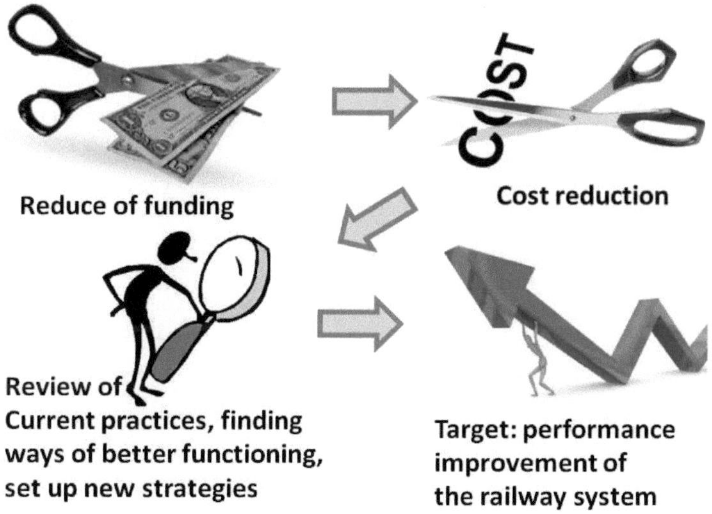

Fig. 15 The today requirements of the external environment to the Infrastructure Manager

K. Tzanakakis: *The Railway Track and Its Long Term Behaviour*, STTT 2, pp. 29–30.
DOI: 10.1007/978-3-642-36051-0_9

The rail Infrastructure Manager (IM), just like any manager, has to efficiently make use of the means made available to him. The IM however, has to deal with a complex situation in terms of high and difficult to control cost structures, and the often highly unstable funds to be collected from both public and private actors. It is necessary to find the optimal balance between reducing these costs on the one hand and on the other to avoid breakdowns that may harm performance and safety. The challenge is to manage cost and revenue as a business and at the same time to take into account public service considerations.

So, the Infrastructure Manager faces the following challenges, to be fulfilled all at the same time:

- Ensure the safe operation of the railway infrastructure
- Decrease his costs for the maintenance of the railway infrastructure and the traffic management
- Increase the efficiency of the railway infrastructure.

The Infrastructure Manager has to optimize his way of operation. The only way to achieve this target for the maintenance of railway infrastructure is included in the word "**Quality**". If the Infrastructure Manager can achieve to get a new track of high quality, to maintain it with the optimum maintenance strategies, he will be able to manage a track with low costs and a high efficiency.

10 Multi-Annual Contracts

Multi-annual contracts represent a long-term financing arrangement between the State and the rail Infrastructure Manager for infrastructure maintenance. Multi-annual contracts should force both parties to take a long-term view and develop maintenance plans on the basis of the infrastructure manager's business plan and thus on future service demand.

The table below sets out maintenance cost reduction potentials as quantified by Infrastructure Managers and transport ministries in 2007[1]:

Table 2 Maintenance cost savings[2]

Multi-annual contract will reduce maintenance cost because of	*Expected cost reduction*
• **more efficient use of resources**	2 – 5%
• **increased efficiency in outsourcing maintenance**	5-10 %
• **more advanced personnel reduction policies**	0,1- 3 %

Under the pressure of 'use it or lose it', maintenance activities tend to peak towards the end of the year. The logic of public budgeting is such that budgets tend to be cut if they are not fully used in previous years. Massive maintenance works at this time also cause more delays and disrupt the service. Abandoning yearly planning in favour of multi-annual schemes thus reduces overall disruption as maintenance works can be planned such that traffic is disrupted as little as possible.

Multi-annual contracts are recommended to improve the quality and maintenance of infrastructures in this sector. Current European Union (EU) legislation requires defining measures to reduce the costs of infrastructure provision and of charges for their use, taking into account safety and maintenance of the quality of the infrastructure service.

The availability and the quality of the infrastructure have a strong impact on the competitiveness of the rail sector. However, maintenance of infrastructure does not always gain the finance that railway operators expect to enable them to compete with other modes of transport. Almost one third of managers state that the finance available to them is not sufficient to maintain their network.

If properly negotiated and prepared, a multi-annual contract can bring many advantages. More particularly, its role is to:

---

1 Communication from the Commission to the Council and the European Parliament - Multi-annual contracts for rail infrastructure quality -COM/2008/0054 final.

2 Source: EU rail infrastructure managers, PriceWaterhouseCoopers 2007.

K. Tzanakakis: *The Railway Track and Its Long Term Behaviour*, STTT 2, pp. 31–32.
DOI: 10.1007/978-3-642-36051-0_10

- **Provide a long-term financing framework for maintenance** forcing both parties to take a long-term view and develop maintenance programs on the basis of future service demand. In fact, it is important that the rail infrastructure corresponds with future transport demand structures in order to boost traffic and revenues. These contracts also permit trade-offs between taxpayers' and users' interests, between maintenance and quality of the network, and between short-term maintenance and renewal;
- **Complement the charging system** by transfers made within the framework of these contracts so as to ensure financial stability. A multi-annual contract has to be consistent with the charging framework, which has to comply with the existing charging rules;
- **Enable effective cost control** by long-term planning of rail maintenance to reduce unit costs. This procedure enables the volume of work to be adapted without changing plans at the last minute. With a multi-annual allocation, the manager can actually make use of funds in a more flexible manner which is thus better suited to business needs, rather than according to the rigid rules of public spending;
- **Enable benchmarking and regulatory supervision** by setting performance targets more effectively. In fact, setting more precise performance targets makes it easier to gauge the relative positions of Infrastructure Managers and to define cost effectiveness in terms of a national infrastructure manager's cost elements and their performance in comparison with other managers;
- **Improve performance** based on performance-related payments and not on compensating the Infrastructure Manager for a particular expenditure as well as reinforcing **quality control**. Quality criteria can be divided into two categories: Indicators based on the quality of the train service (i.e. speed, safety) and indicators based on infrastructure provision (i.e. maintenance costs per km of track, percentage of lines under temporary speed restrictions);
- **Secure the effectiveness of contractual agreements** providing for example, sanctions applicable in cases of non-compliance. The monitoring process should be undertaken by an independent body rather than by the two contracting parties. Sanctions may consist of penalties (fines), reduced financial input or even a replacement of Infrastructure Managers and need to be progressive and in proportion with the infringement.

EU Member States and their Infrastructure Managers should conclude multi-annual contracts which comply with the national strategic transport plan and with the Infrastructure Managers' business plans. The State should consult stakeholders on any proposal for multi-annual contracts before entering into a new contract or renegotiating existing provisions.

Infrastructure Managers should check track condition and should indicate cases where infrastructure quality is considered to be substandard.

Part C

Functional Knowledge of the Railway Track System

Knowledge is power

Francis Bacon

Section III: The Railway Track as a System

"An important aspect of a part's performance is how it interacts with other parts to affect the performance of the whole."
Russell Ackof

In this Section the knowledge about the "how" and "why" of the track behaviour is provided, as a base knowledge for highlighting the importance of track quality.

11 Introduction

A "system" can be defined as a complex of elements standing in interaction.

Systems thinking involves inter alias shifting attention

- from the parts to the whole,
- from objects to relationships,
- from structures to processes

When we use systems thinking, we are actively asking ourselves "**Why is this working as it does**?"

The railway system is a very complex system: it is often a mixture of components of different age and status that have to work together in a system. The railway track has to guide the trains in a safe and economic manner. The track and the switches should allow smooth passage of the trains [24][24.1].

If the track is not perfectly levelled and aligned, the irregularities will cause oscillations or vibrations of the train, and this may bring discomfort for passengers and damage for goods.

Long wavelength irregularities of the track will give rise to low-frequency oscillations of the train. Short wavelength irregularities cause vibrations and noise, both in the train and in the environment.

Oscillations, vibrations, and noise may become unpleasant for the train passengers and for people close to the railway line.

For the track engineer it is important to know "how" the track functions, but also "why" it functions in a specific manner, in order to be able to predict the effect of changes (typical cases of changes are ex. running with higher speeds, increasing the density of traffic, change of rail type –profile, hardness- etc.).

K. Tzanakakis: *The Railway Track and Its Long Term Behaviour*, STTT 2, p. 37.
DOI: 10.1007/978-3-642-36051-0_11

12 Forces on the Track

Key terms:
Forces, quasi-static forces, dynamic forces, P1 force, P2 force, impact forces

12.1 Overview

Both vehicle and track have irregularities

- track geometry irregularities (long wavelength) or
- discrete irregularities (short wavelength) in the wheel tread or rail running surface.

These two classifications of track irregularities (long/short wavelength) produce different magnitudes of forces due to the resonances they create within the track structure.

In order to understand the damage mechanism of the track and the deterioration of the track geometry, it is necessary to recognize all the forces to the railway track from a train running above it and the responses made to these forces.

It is also important to fully understand the causes of these induced forces in order to be able to remove them from the track and therefore extend the life of the track asset.

The forces acting on the track result from the train loads (there are also thermal loads, which isn't the subject of this chapter).

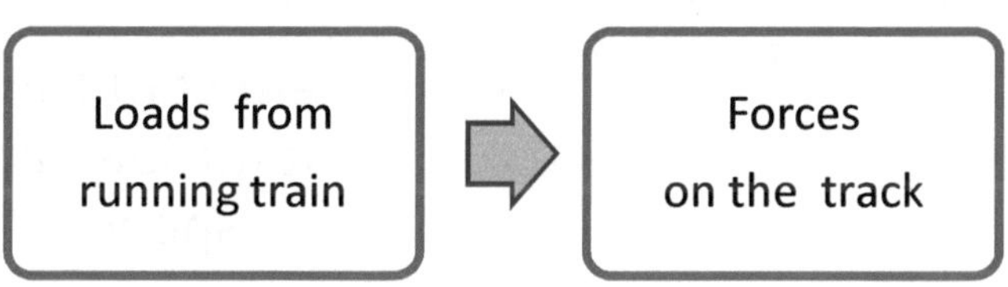

Those loads can be categorised into three main groups:

- **Vertical** loads
- **Lateral** loads (transverse to the track), and
- **Longitudinal** loads (parallel to the track).

K. Tzanakakis: *The Railway Track and Its Long Term Behaviour*, STTT 2, pp. 39–47.
DOI: 10.1007/978-3-642-36051-0_12

Depending on their nature, those loads can be divided into three groups:

- **static loads** (normally caused by the vehicle body mass)
- **Quasi-static loads (or dynamic ride loads),** which are associated with vehicle movements: Vehicles running on the track apply certain forces on the track structure due to the behaviour of the vehicle body, bogie and other masses in reaction to geometrical irregularities in the track. These are low frequency forces (see par. 12.2) and
- **Dynamic (dynamic wheel/rail) loads,** which are associated with significant irregularities that may occur during the life of the track structure and vehicle [61] (see par. 12.3):
    - Irregularities of the track geometry
    - Discontinuities on the running surface (switches, joints)
    - Wear of the running surface of the rails
    - Wear of the wheels (out-of-round wheels)
    - Vehicle suspension and vehicle asymmetries

  Dynamic wheel/rail forces are much higher in magnitude than quasi-static (dynamic ride) forces.

In this chapter we'll concentrate on quasi-static and dynamic forces.

12.2 Quasi-static (Dynamic Ride) Forces

The quasi-static load can be defined as the sum of the static load and the effect of the static load at speed[3] and they are classified in the frequency range between 0,5 and 30 Hz [9].

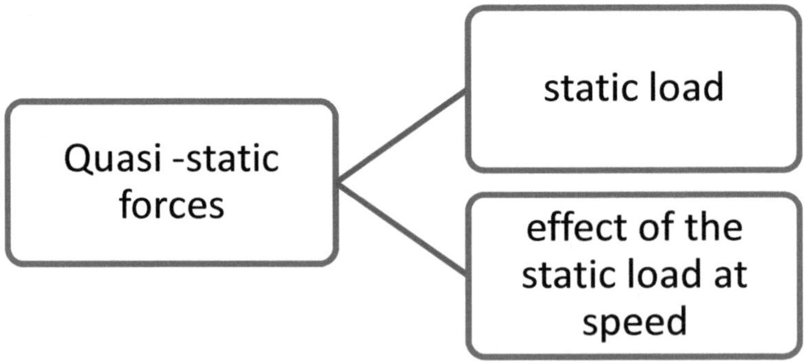

The load includes the effects of

- the geometrical roughness of the track on vehicle response and
- unbalanced superelevation (the effect of the train load not being distributed evenly over both rails).

[3] See the Australian Standard for pre-stressed concrete sleepers (AS1085.14, 2003).

If both the wheel tread and rail surface are in good condition, the wheel/rail contact force would be similar to the static wheel load.

The quasi-static force has been found to be typically between 1,4 and 1,6 times the static wheel load before unbalanced superelevation effects are included[4]. Forces greater than 1,6 times the static wheel load are found to be the result of more significant wheel/rail contact conditions.

12.3 Dynamic Forces

12.3.1 The Dynamic Wheel / Rail Forces

Dynamic forces come in two categories:

- **P1 Forces**
- **P2 forces**

In the following paragraphs a more analytical presentation of the P1, P2 forces is given [60].

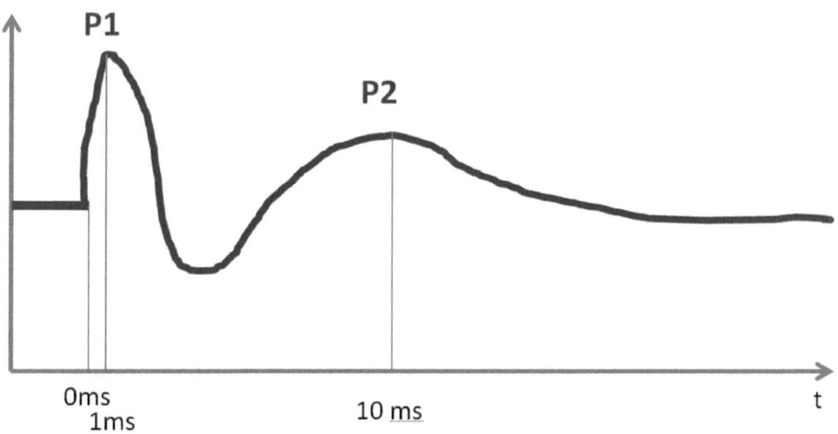

Fig. 16 Vertical forces after a dipped rail joint (after [59])

12.3.2 The P1 Force

P1 forces are classified in frequency range between 100 Hz and 2000 Hz. These forces are also called **impact forces**. They correspond to **surface irregularities** or **defects in rails and wheels** and produce strong impact to rail and wheel (Figure 17).

P1 is a very high frequency force occurring ¼ - ½ ms after crossing the angular discontinuity that occurs at the bottom of the dip and has a very short duration

[4] As above.

(Figure 16). At that point the rail and sleepers due to their inertia resist the downward motion of the wheel and the contact zone between wheel and rail is compressed.

The P1 force is responsible for rail hammering just after the joint gap and produces high stresses in the rail web. It contributes to bolt hole failures in bolted joints by increasing the stress range. These forces contribute to the cracking of concrete sleepers[5]. Its effects are largely filtered out by the rail and sleepers, and do not directly affect ballast or subgrade settlement [9]. However, they have a great influence on wheel/rail contact behaviour.

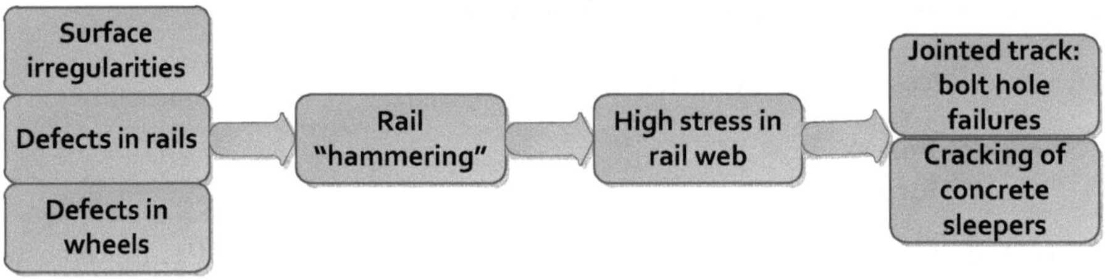

Fig. 17 Consequences of P1 forces

12.3.3 The P2 Force

P2 forces are classified in the frequency range between 30 Hz and 100 Hz (lower frequency range). They correspond to the movement of unsprung masses of the vehicles (see par.14.2.2).

The P2 forces have a lower-amplitude and longer-duration than the P1 forces.

The peak force occurs in the area of the first running-off sleeper after the joint. The wheel set mass and the rail/sleeper mass move down together and compress the ballast beneath the sleeper. P2 forces therefore increase contact stresses, contribute to the total stress range experienced by the rail web and at joint bolt holes and also increase the loads on sleepers and ballast in the immediate neighbourhood of the joint. For this reason P2 forces are of great interest to the track design engineer.

The P2 forces contribute primarily to the degradation of track geometry. As shown in Figure 16, the P1 impact forces could be as high as 3,5 times the static load imposed by the wheel on the rail, and the P2 impact forces could be as high as 2,5 the static load.

12.3.4 Synopsis

The following table gives a better overview of the above described forces on the track:

[5] Experienced by Amtrak in the late 1970s and early 1980s [59].

Table 3 The dynamic forces on the track

Force	Cause	Mechanism	Impact	Frequency area
Quasi-static forces	• gross mass of vehicle, • movement of sprung mass of the vehicle, • centrifugal force in curves, • cross winds			0,5 - 30 Hz
P1 Forces (also called impact forces)	surface irregularities or defects in rails and wheels	At the bottom of the dip, the rail and sleepers resist the downward motion of the wheel and the contact zone between wheel and rail is compressed.	• strong impact to rail and wheel • relevant to track deterioration	100 - 2000 Hz
P2 forces	unsprung masses of the vehicles	The wheel set mass and the rail/sleeper mass move down together and compress the ballast beneath the sleeper.	• relevant to track deterioration	30 - 100 Hz

12.4 Force Limits

Railway organizations around the world have set limits on the various forces at the wheel / rail contact area [59].

12.4.1 Vertical Force Limits

In the vertical direction high forces can cause damage to the rails and supporting structures and can cause rolling contact fatigue when combined with high tangential forces such as occur during traction, braking or curving.

UIC Leaflet 518 ([70]) sets a maximum static load of 112,5 kN per wheel and a maximum dynamic vertical force per wheel of between 160 kN and 200 kN, depending on maximum speed (provided this values does not exceed the static wheel load plus 90 kN). In small radius curves (less than 600 m) UIC Leaflet 518 sets a limit of 145 kN for the quasi-static vertical force.

12.4.2 Lateral Force Limits

In the lateral direction high forces can cause distortion of the track on ballast-bed. This is normally protected against by using the simple but widely established

PrudHomme limit for the track shifting force at one wheel set, which can be calculated from the static load (P_0 force):

$$Y = 10 + \frac{P_0}{3} \quad \text{for passenger trains}$$

$$Y = 0{,}85 \cdot \left(10 + \frac{P_0}{3}\right) \quad \text{for freight trains}$$

where Y and P_0 are in kN.

Lateral forces of very short duration are less likely to shift the track and therefore only forces that act for more than 2 m of track length are usually counted. In small radius curves (less than 600 m) UIC Leaflet 518 sets a limit of 60 kN for the quasi-static lateral force.

12.4.3 Derailment

Possibility of wheel climb derailment is indicated by the ratio of the lateral force Y to vertical force V as mentioned earlier. The theory of Nadal is used to establish limits for the Y/V derailment ratio with 0,8 as the limiting value in UIC Leaflet 518.

$$\frac{Y}{V} \leq 0{,}8 \quad \text{or} \quad Y \leq 80\%\, V$$

12.4.4 Wheel Unloading

Very low vertical forces at the contact patch can indicate that a vehicle is tending to derailment by rolling over or by failing to follow twists in the track. In the United Kingdom a lower limit of 60% of the static wheel load (i.e., unloading by over 40%) is set [59].

12.5 Synopsis

The following table gives a synopsis of the vertical forces, their impact on the track, the diagnosis of the impacts and remedy measures.

Table 4 Vertical Track Forces (Shelley and Williams 1990) (after [64])

CAUSE	FORCE	SYMPTOM	DIAGNOSIS	REMEDY
Impact at Rail Welds	Rail: P1 + P2	• Rail fatigue failure • Corrugations • Pad degradation • Sleeper cracking/movement • Ballast degradation • Weld fatigue	• Ultrasonic inspections • Manual inspections	• Weld straightening • Rail grinding • Lower unsprung mass • Lighter axle loads
Vehicle/Track Interaction	Quasi Static Dynamic Forces	• Track geometry deterioration • Rail failure/fatigue • Ballast failure/degradation • Subgrade failure/degradation	• Track recording car • Ultrasonic inspections • Manual inspections	• Tamping • Mud hole rectification • Ballast cleaning • Ballast regulating • Formation/ballast design
Wheel irregularities	Wheel: P1 + P2	• Sleeper cracking • Rail breaks Wheel cracks • Ballast deterioration	• Wheel impact detector	• Wheel turning program • More frequent inspections
Static load	Po	• All of the above	• Weighbridge • Manual checking	• Lighter axle loads • No over loading

12.6 The Influence of Vehicle and Track Characteristics to the Forces on the Track

The magnitude of the track forces affects directly the magnitude of track degradation. In the following, some aspects determining forces and their influence on track degradation are presented.

12.6.1 Vehicle Characteristics

The vehicle characteristics such as speed, axle load, unsprung mass, suspension, wheel profile, axle spacing etc. influence the magnitude of track forces (static, quasi-static, and dynamic). The current condition of the wheel such as wheel flats and wheel corrugation contributes to dynamic forces.

12.6.2 Track Characteristics

Track design **geometry** (curves etc.) influences static and quasi-static forces.
Track geometry **quality** is a main contributor to dynamic forces.
Corrugation[6] generates high-frequency forces.
Rail imperfections such as joints or poor welds lead to impact forces.

12.7 Synopsis on the Influence of Vehicle and Track Characteristics to the Forces on the Track

The following table gives a synopsis of the influence of vehicle and track characteristics to forces on the track.

Table 5 Influence of vehicle and track characteristics to forces on the track

System	System characteristics	Influence on the track forces
Vehicle	• speed • axle load • unsprung mass • suspension, • wheel profile • axle spacings	• static • quasi-static • dynamic forces
	condition of the wheel	dynamic forces

[6] Often called rail roughness.

Table 5 (*continued*)

Track	Track design geometry	static and quasi-static forces
	Track geometry quality	dynamic forces
	Corrugation	high-frequency forces
	Rail imperfections	impact forces

13 The Track Subsystem

Key terms:
Train guidance, wheel/rail, loads, track components, rails, fasteners, Rail pads, sleepers, ballast, Sub-ballast, subgrade

The railway track has to fulfil two main functions:

- to **guide the train** with safety
- to **carry the load** of the train and to **distribute** the load to the subgrade over an area that is as large as possible.

In this Chapter we will examine those two functions, and also the components of a track, which are important elements for the proper track function.

13.1 Functions of the Track

13.1.1 Train Guidance

The main function of the track is to **guide the train** with safety.

The two wheels of a wheel set are rigidly connected to the wheel axle. The wheel profiles are conical in order to steer the wheel set. Normally, the wheel set is not exactly on the centreline of the track. Due to this eccentricity and the conicity of the wheel set, one wheel will have a larger rolling radius than the other.

The motion of the wheel set is described in the next table.

Table 6 The motion of the wheel set

Case A: The wheel set exactly at the centreline. The rolling diameter of the left wheel (D_{left}) is equal to that of the right wheel (D_{right}). This situation is only theoretically possible.	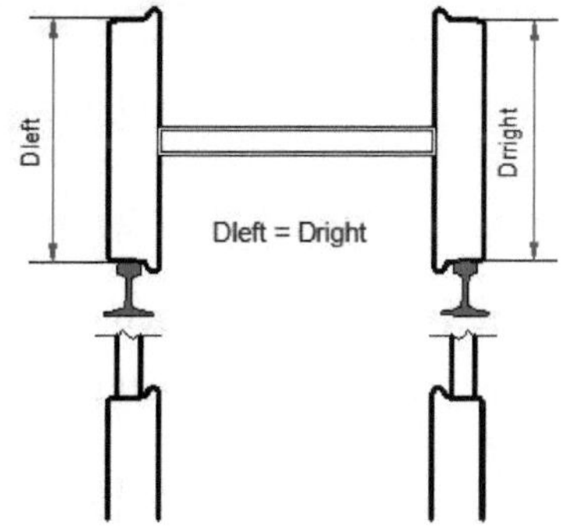

K. Tzanakakis: *The Railway Track and Its Long Term Behaviour*, STTT 2, pp. 49–58.
DOI: 10.1007/978-3-642-36051-0_13

Table 6 (*continued*)

<table>
<tr><td><u>Case B: The wheel set on a right curve (wheel flange position to the left railhead)</u>
The rolling diameter of the left wheel (D_{left}) is greater than to that of the right wheel (D_{right}).
The left wheel with the larger rolling radius forces the wheel set back towards the centre.</td><td>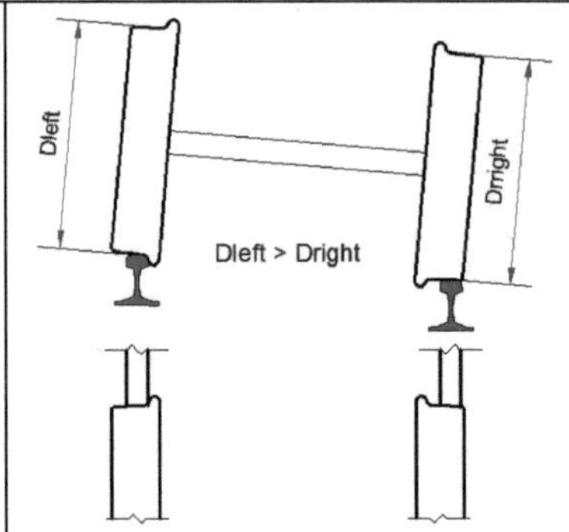</td></tr>
<tr><td><u>Case C: The wheel set on a left curve (wheel flange position to the right railhead)</u>
The rolling diameter of the left wheel (D_{left}) is less than to that of the right wheel (D_{right}).
The right wheel with the larger rolling radius forces the wheel set back towards the centre.</td><td>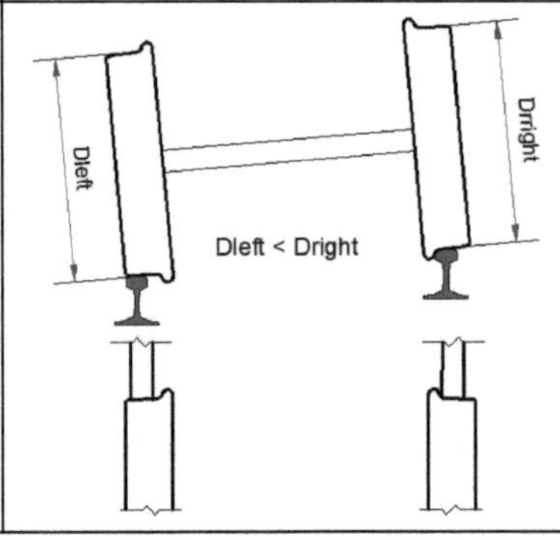</td></tr>
</table>

The motion of the wheel set described, leads to a sinusoidal movement of the wheel set along the track (Figure 18). This movement will induce a low-frequency lateral movement of the railway vehicle and generate lateral forces on the track.

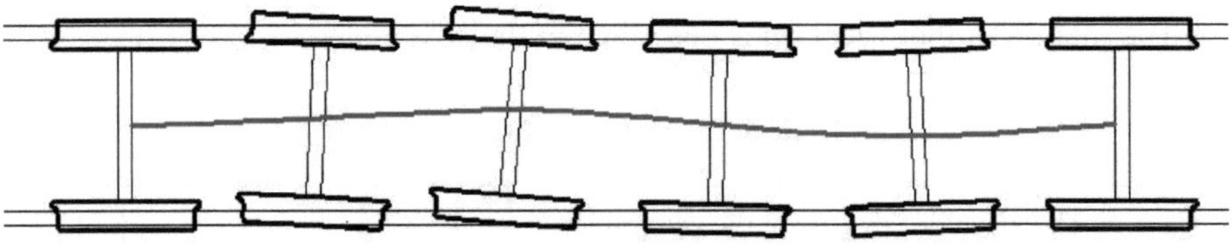

Fig. 18 Sinusoidal motion of wheel set

13.1.2 Wheel/Rail Contact Profiles

Figure 19, shows the different contact profiles in

- a straight line (left profile),
- a curve
    - at the outer "high rail" which is situated higher in a curve due to canting, when the wheel flange contacts the railhead (the train

runs with the full operation speed, for which the superelevation is given) (middle profile), and
- o at the inner "low rail" which is lower than the outer rail, when the wheel flange is away from the railhead (right profile).

The inside of the rail head facing the other rail is called the "gauge face" and the outside of the rail head is, for obvious reasons, called the "field face".

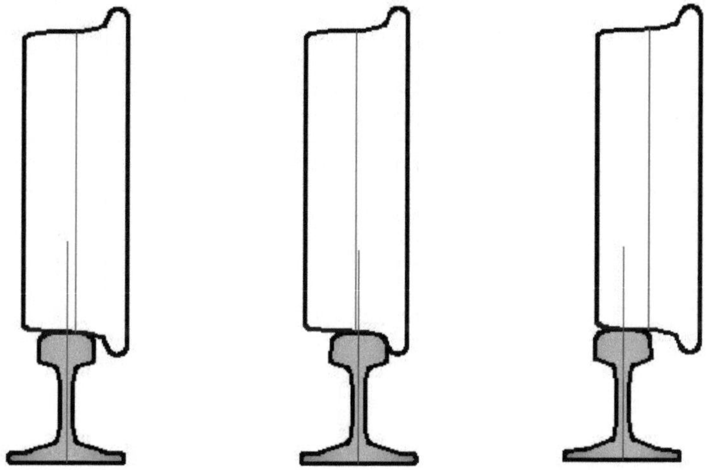

Fig. 19 Wheel/Rail contact profiles (left: straight line, middle: "high rail" in a curve, right: "low rail" in a curve)

The wheel / rail interface will be discussed in Chapter 17.

13.1.3 Carrying the Loads of the Trains

Another function of the track is to **carry the load** of the train and to **distribute** the load to the subgrade over an area that is as large as possible (see Figure 20).

At the contact area between the wheel and the rail, there is a high pressure (in the figure: a force of Q=125 KN → p = 42000 N/cm^2). This pressure is reduced gradually in the system components, so that the pressure to the subgrade has a value of about 10-12 N/cm^2, which is about 0,023% of the pressure at the wheel-rail contact area, or a reduction of 3400-4200 times. The following table shows, the gradually reduction in the system components.

Table 7 The gradually reduction of the pressure due to load in the system components

	Area A	Pressure p		Reduction	
				Between neighbouring components	Between Wheel and current component
Wheel-Rail contact	3 cm^2	125000 N /3 cm^2 =	42.000 N/cm^2	-	-
Rail - foot	200 cm^2	125000 N /200 cm^2 =	625,00 N/cm^2	67	67
Rail seat	510 cm^2	125000 N /510 cm^2 =	245,10 N/cm^2	3	170
Bottom of a sleeper	2380 cm^2	125000 N /2380 cm^2 =	52,52 N/cm^2	5	793
Bottom of ballast area	10100 cm^2	125000 N /10100 cm^2 =	12,38 N/cm^2	4	3367

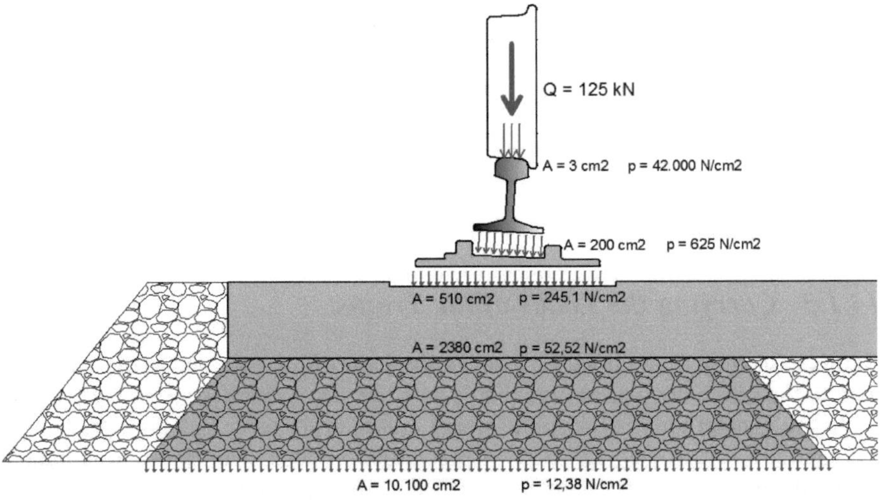

Fig. 20 Pressure distribution of the wheel force Q in the individual system components of the track

13.2 The Track Components

Track components are grouped into two main categories (Figure 21),

- the **superstructure** which consists of the
  - rails,
  - sleepers and

- a fastening system to hold the components together
- ballast[7], and

- the **substructure**, which consists of the
  - sub-ballast and the
  - subgrade or formation.

The other elements are rail joints or welds in continuous welded rail (CWR).

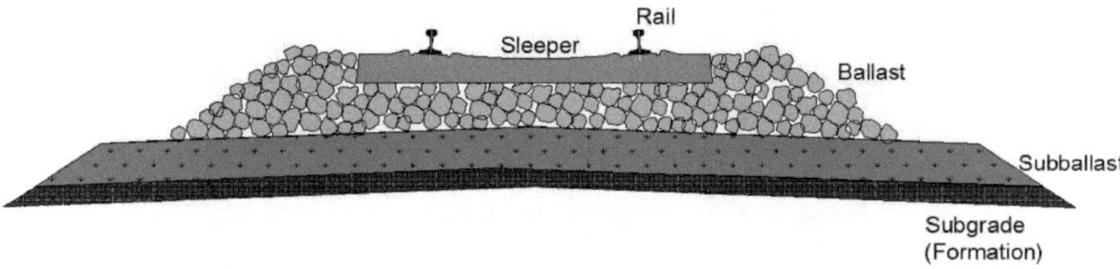

Fig. 21 Typical ballasted track system

The typical track structure is ballasted track. We will here be focusing on the ballasted track structure.

13.3 The Characteristics of the Track Components

13.3.1 Rails

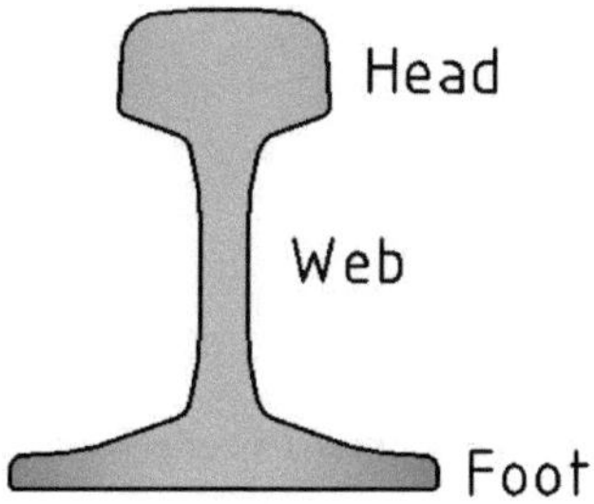

Description	**Rails** are the longitudinal steel members that directly guide the train wheels evenly and continuously. They must have sufficient stiffness to serve as beams that transfer the concentrated wheel loads to the spaced sleeper supports without excessive deflection between supports ([27]).

[7] In some references "ballast" is counting to the substructure.

13.3.2 Fasteners

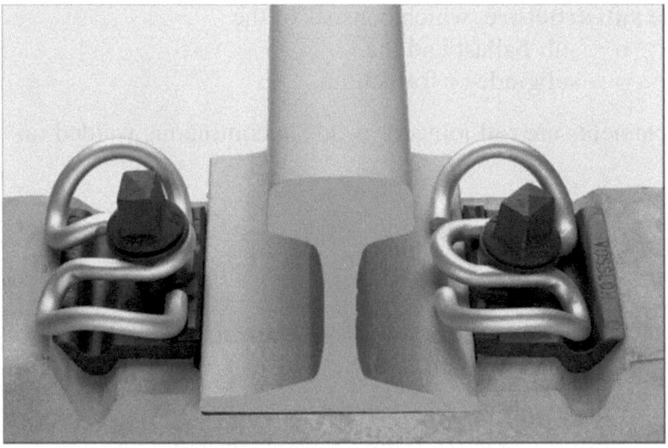

Fig. 22 Fastener W14 ("Skl 14") - Vossloh

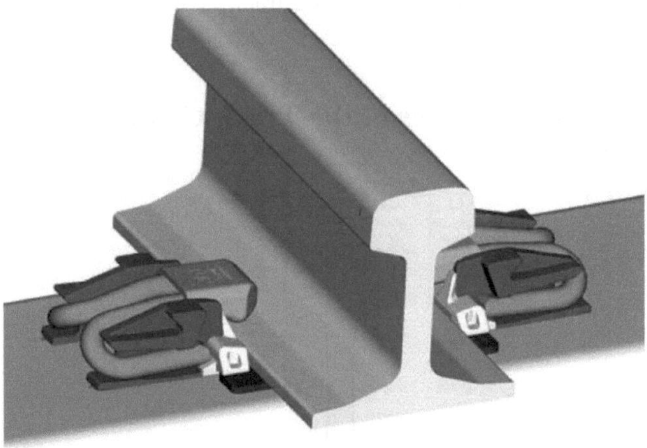

Fig. 23 Fastener "Fastclip" - Pandrol

Key Features	A fastener must meet the following criteria [30]: • The longitudinal resistance of the fastener must be greater than that of the longitudinal resistance between sleeper-ballast. • There must exist a safety factor for the longitudinal resistance, able to cover the longitudinally unequal distribution of clamping forces, dynamic phenomena, etc. • In case of rail rupture, the resulting gap should remain small.

Function	Fasteners are typically required to retain the rails on the sleepers, to maintain the track gauge and to resist vertical, lateral, longitudinal and overturning movements of the rails. Causing these movements are forces from the wheels and from temperature change in the rails ([27]).

13.3.3 Rail Pads

Key features	The modern rail pads –taking into account the today's environment where the axle loads and speeds are being increased-shall: 1. Reduce vibration and impact transmission from rail to sleeper by providing resilience and impact attenuation. 2. Give adequate resistance to longitudinal and rotational movement of rail. 3. Ensure long life - minimum 10 years. 4. Ensure that the rail pad properties are stable over a wide range of operating temperatures during life. 5. It shall not deform, abrade or move out of fastening system under dynamic loads. 6. It is shall not be adversely affected by ozone, ultra violet light, oils and related chemicals. 7. Reduce possibility of rail foot corrosion and concrete sleeper erosion 8. Provide more resistance to longitudinal rail movement due to creep or thermal expansion. 9. Provide electrical resistance between rail and sleeper 10. Provide conforming layer between rail and sleeper to avoid contact areas of high pressure.

Function	Rail pads are required between the rail seat and the sleeper surface to fulfil various functions. From a track dynamics point of view, the Rail pads play an important role. They influence the overall track stiffness: • When the track is loaded by the train, a soft Rail pad permits a larger deflection of the rails and the axle load from the train is distributed over more sleepers. • Soft rail pads isolate high-frequency vibrations. They suppress the transmission of high-frequency vibrations down to the sleepers and further down into the ballast[8].

13.3.4 Sleepers

Description	Sleepers are essentially beams that span across and tie together the two rails.
Function	The sleepers (among others) • receive the load from the rail and distributes it over the supporting ballast at an acceptable ballast pressure level • hold the fastening system to maintain proper track gauge • restrain the lateral, longitudinal and vertical rail movement by anchorage of the superstructure in the ballast • provide a cant to the rails to help develop proper rail-wheel contact by matching the inclination of the conical wheel shape.

[8] A stiff rail pad gives a more direct transmission of the axle load, including the high-frequency load variations, down to the sleepers below the wheels.

13.3.5 Ballast

Description	Ballast is the layer of crushed stone on which the sleepers rest.
Function	The ballast (among others) • **distributes load** by from the sleepers uniformly over the subgrade assisting in track stability • assists in **absorbing shock from dynamic loads** by having only a limited spring-like action (due to the rough interlocking particles) • **anchors the track** in place against lateral, vertical and longitudinal movement by way of irregular shaped ballast particles that interlock with each other • **easily drains** any moisture introduced into the system through the ballast away from the rails and sleepers. • assists in **track maintenance** operations due to its easy manipulation.

13.3.6 Sub-ballast

Description	Sub-ballast is material chosen as a transition layer between the upper layer of large-particle, good quality ballast and the lower layer of fine-graded subgrade.
Function	Sub-ballast • assists in reducing the stress at the bottom of the ballast layer to a tolerable level for the top of the subgrade • can prevent the inter penetration of the subgrade and ballast, thereby reducing migration of fine material into the ballast which effects drainage • acts as a surface to shed water away from the subgrade into drainage along the side of the track.

The usual thickness of the gravel sub-ballast layer is 15 cm. However, some railways do not use a sub-ballast layer and they simply use a greater thickness of the subgrade (formation) layer.

13.3.7 Subgrade (or Formation)

Description	Subgrade, or formation, is a surface of earth or rock levelled off to receive a foundation for the track bed. Sometimes an extra layer, a formation layer, is put on the earth so as to give the correct profile of the track bed. On this material the sub-ballast and ballast layers rest.
Function	Subgrade • offers the final support to the track structure • bears and distributes the resultant load from the train vehicle through the track structure • facilitates drainage and provides a smooth platform, at an established grade, for the track structure to rest upon.

14 The Vehicle Subsystem

> *Key terms:*
> *Car body, bogie, unsprung mass, primary suspension, secondary suspension, wheel, wheel set*

14.1 Overview

Today there is a general trend towards increased train speeds (see also SECTION II). However, higher speeds usually generate increased forces and accelerations on the vehicle and subsequently forces on the track. This leads to requirements for lighter car bodies, which reduce the impact between wheels and rails. However, lowering the car-body weight also means a reduction of the stiffness of the vehicle structure, which results in lower natural frequencies. This increases the risk of resonance vibrations, which negatively affects ride comfort. Car body vibrations can be reduced either by focussing on the structural stiffness of the system or by optimizing the damping components ([56]).

Trains (or rolling stock) consist of two types of vehicle (Figure 24):

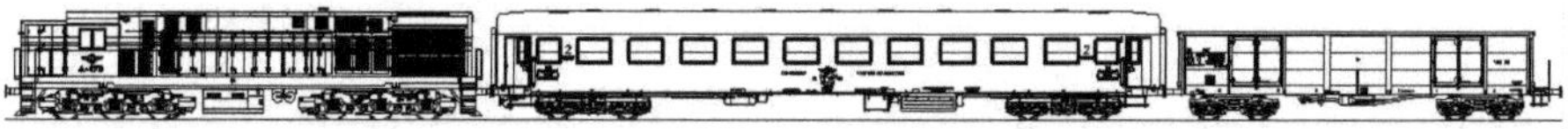

Fig. 24 Train composition of a locomotive, a passenger coach and a freight wagon

- a locomotive or power-car that enables the train to move, and
- wagons
    - passenger coaches that carry passengers
    - freight wagons that carry goods (freight).

To investigate rail track forces, vehicles can be defined by the number of axles they have: locomotives typically have triple axle bogies; wagons may have triple, double or single axle bogies.

K. Tzanakakis: *The Railway Track and Its Long Term Behaviour*, STTT 2, pp. 59–63.
DOI: 10.1007/978-3-642-36051-0_14

14.2 Structure of a Typical Vehicle

A typical vehicle consists of a car body and bogies (Figure 25).

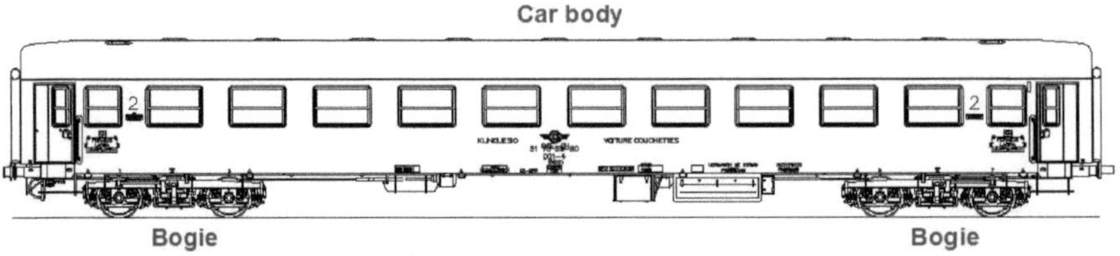

Fig. 25 A typical passenger coach

14.2.1 The Car Body

The **car body** is best defined as a "container" for the passengers or freight. The vibrations of a car body are mainly caused by track irregularities that are transmitted up to the car body via the bogies.

14.2.2 The Bogie

The **bogie** is the part that guides the train on the rails and provides stable operation. There are two main types of bogies, one with and the other without primary suspension (see 14.2.3).

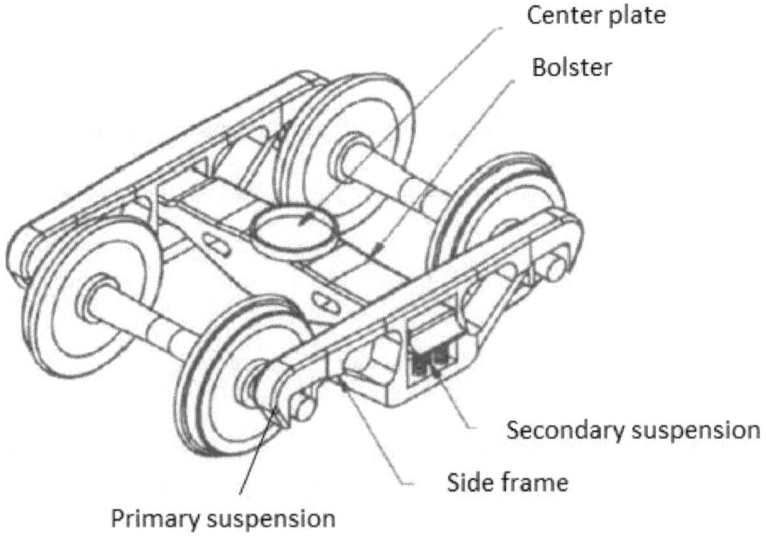

Fig. 26 Three piece bogie ([9])

The suspension is a very important element of the bogie, and determines the unsprung masses. The **unsprung mass** is the mass of a wheel, or wheel set, and other associated components which are not dynamically isolated from the track by vehicle suspension arrangements.

Figure 26 shows a typical three piece bogie. Load from the vehicle is applied through the **centre plate** or bowl, where longitudinal or lateral relative movement is restrained. The load is then transferred to the **bolster** which spans between the two side frames and rest on secondary suspension. The **secondary suspension** then takes the load and transfers it to the **side frame** ([9]).

Figure 27 shows the bogie of a modern train.

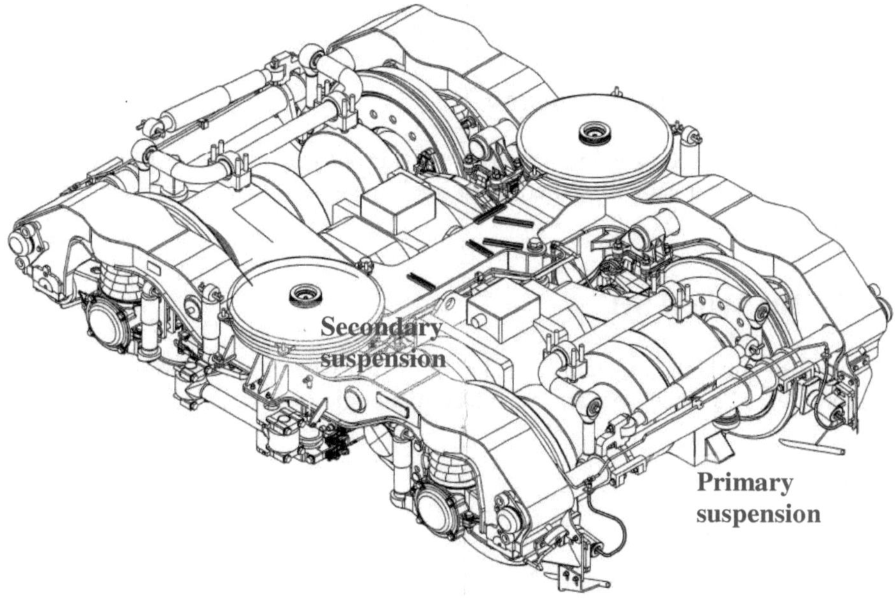

Fig. 27 Boogie of a modern train

14.2.3 Primary Suspension

The primary suspension consists of spring and damper components between the bogie and the wheel set, in order to secure a stable running behaviour, and also to ensure low track forces, low wear and good behaviour in curves.

By including primary suspension the, lateral stability of the wagons and the ride quality can be significantly improved. This is due to the lower unsprung mass as the size of the unsprung mass directly relates the amount of force transferred into the car body and track from wheel or track irregularities.

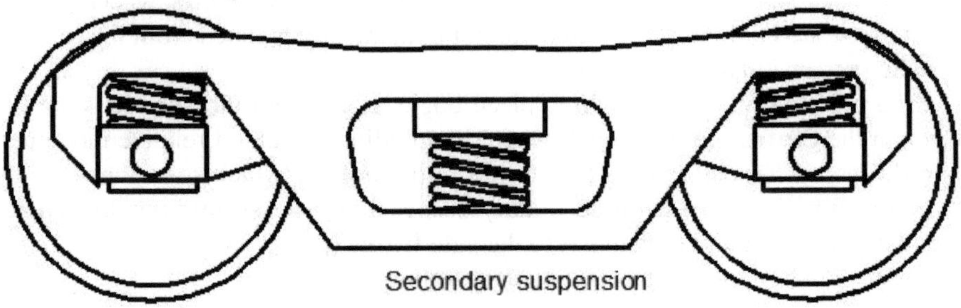

Fig. 28 Typical bogie suspensions (simple bogie)

14.2.4 Secondary Suspension

The secondary suspension interconnects the car body and bogie. It isolates the car body from excitations transmitted from track irregularities via the wheel sets and the bogie frames.

14.3 The Wheel Set

The **Wheel set** is the assembly consisting of two wheels and bearings on an axle (Figure 29). The wheel sets are connected to a bogie.

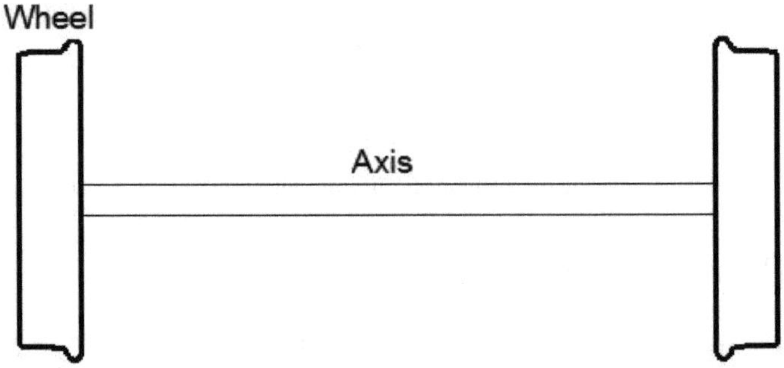

Fig. 29 Wheel set

14.4 The Wheel

The **wheel** is the contact element connecting the vehicle to the track (Figure 30). The conical form of the wheels helps the wheel set along straight or tangent track and slight lateral displacements of the track. The conical shape also assists steering through curves due to the rolling radius difference generated by the lateral displacement of the wheel set in curved track (for a more analytical description see par. 13.1.1)

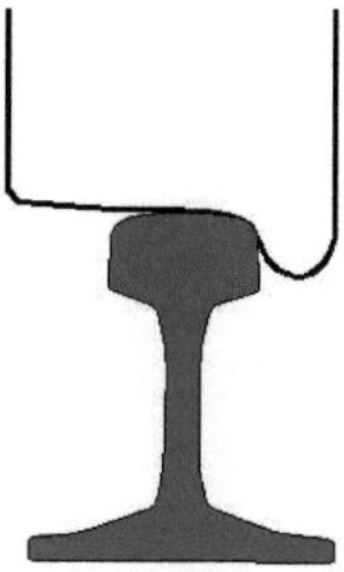

Fig. 30 The wheel as the contact element of the vehicle to the track

15 Modelling the Track

Key terms:
Railway system, dynamic system, mechanical characteristics, track components, rails, fastening system, sleepers, ballast, sub-ballast, subgrade, track dynamic behaviour, continuously supported models, discretely supported rail beam, Beam on Elastic Foundation (BOEF), Euler – Bernoulli beam, Rayleigh – Timoshenko beam, pin-pin vibration, spring – damper system, vehicle – bridge Interaction

15.1 The Railway System as a Dynamic System

The railway system can be considered as an assembly of structural components with specific mechanical properties. They are characterised by their frequency response function which is directed by mass elastic properties. These parameters define the frequencies at which the structure is likely to vibrate [9].

For analysing the track behaviour it's important to understand the mechanical characteristics of the main elements of the rail track when trying to model the railway system, taking into account dynamic phenomena from the interaction between load (from a running train) and the track components[9].

15.2 The Mechanical Characteristics Considered in Modelling Track Components

A discussion of the mechanical characteristics of the track components considered in modelling track is provided below (see also par. 13.3) [9].

15.2.1 The Rails

Rails are linear elements of an infinite length. This allows modelling the rail as beams. Rails have flexural stiffness in vertical and lateral directions and compression stiffness in the longitudinal direction. Rails also have a shear stiffness which is often neglected.

[9] According to Esveld [26], dynamics can be defined as the interaction between load and structure. Track and vehicles can be seen as a structure.

K. Tzanakakis: *The Railway Track and Its Long Term Behaviour*, STTT 2, pp. 65–78.
DOI: 10.1007/978-3-642-36051-0_15

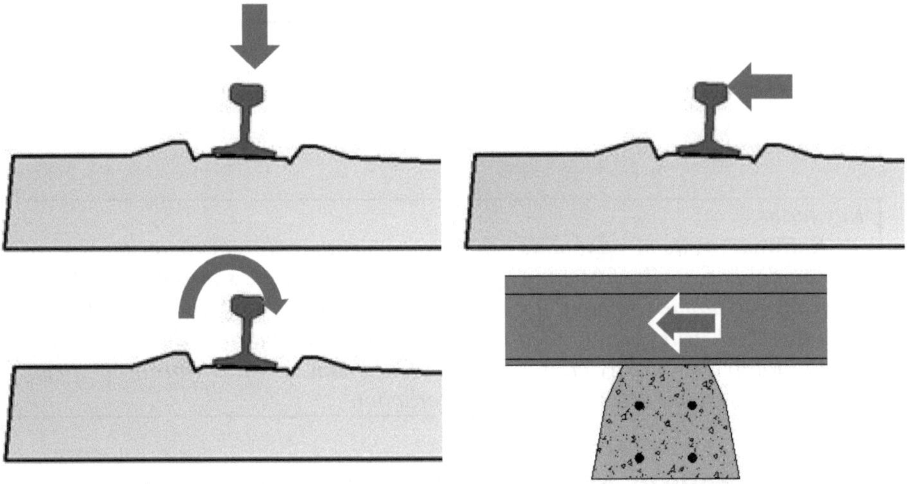

Fig. 31 Vertical, horizontal, rotational and longitudinal forces that the fastening system and rail seat are subjected to under rail vehicle and thermal loading

15.2.2 The Rail Fastening System

15.2.2.1 Overview

The rail fastening system commonly used on concrete sleepers comprises a resilient spring **fastener**, acting essentially in parallel with a much stiffer **rail pad** [9].

The load/deflection behaviour of the fastening system is non-linear ([38]); however, some linearization of the load/deflection behaviour can be justified.

For vertical vibration a pad is usually modelled as a spring and viscous dashpot (damper) in parallel. Rail pads are mainly loaded in compression, permanently by the fastening system by the rail traffic. The inclusion of rail pads is effective in reducing this force as it reduces the effective track mass acting on the sleepers and ballast. Thus **the rail pad is the one component which can be most readily changed to influence the track's dynamic behaviour and rate of degradation**.

15.2.2.2 The Dynamic Interaction between the Rail and the Fastening System

The stiffness of a fastening system is one of the most important characteristics that directly impacts the fastening system's long-term performance under repeated axle loading. The dynamic rail / fastening system interaction can be viewed as a complete set of springs and dampers (see following figures). The stiffness of each component determines how much the rail is allowed to move within the rail seat [69].

(1) Rail - Tension clamp / Upwards motion

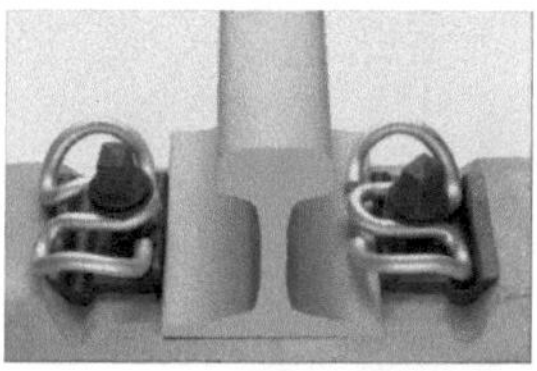

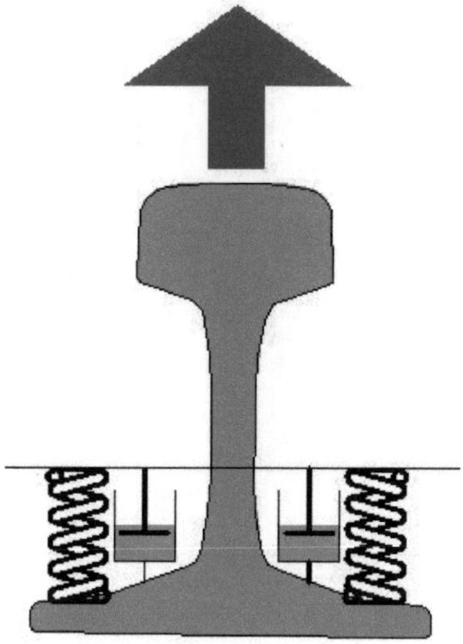

(2) Rail – Rail pad / downwards motion

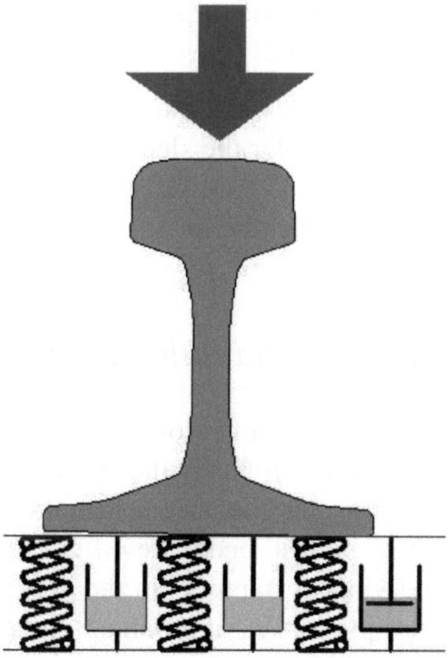

(3) Rail - Tension clamp – Pad / sideward motion	
 	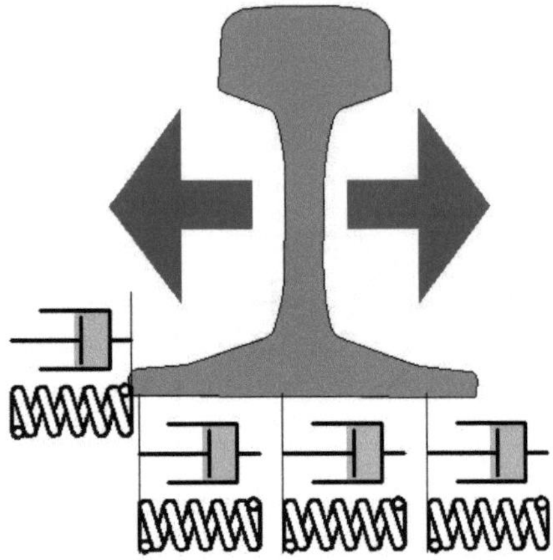

15.2.3 Sleepers

Sleepers may be modelled as

- rigid beams or
- beams with flexural and shear stiffness or
- rigid bodies.

At higher frequencies, the mass of a sleeper becomes increasingly important and it is essential to consider the sleeper as a dynamic component that has both mass and stiffness. Because of its distributed mass and stiffness, the sleeper resonates at a series of frequencies, the most significant for typical sleepers being around 200 Hz and 650 Hz.

15.2.4 Ballast and Sub-ballast

Ballast and sub-ballast is typically modelled as a load distributing material.

Ballast material in particular deflects in a highly non-linear manner under load due to vacuums at the sleepers/ballast interface and also in the ballast itself. Despite this, ballast beds are often modelled by discrete or distributed linear springs and viscous dampers in the vertical direction.

The mass properties of a ballast bed are also important; however the amount of mass that should be incorporated in dynamic models is difficult to estimate.

15.2.5 Subgrade

Subgrade material is similar to ballast material in its properties. The load distribution influences the effective mass and the stiffness and damping properties, so that modelling is considered as very complicated.

15.3 Components to Reduce the Forces

The entire system of railway components can be divided into two subsystems:

- the vehicle
- the track.

Between both subsystems there is a physical wheel/rail contact.

"Suspension systems" are used in order to reduce the forces from the running train to the superstructure and the substructure. The reduction of the forces is done in the vehicle and the track (through suspension systems) as described in Table 8 [9]:

Table 8 Reduction of the forces in the railway system

Stages to reduce the forces	Subsystem	Measures
1st stage	Vehicle	Primary suspension: suspension system between the wheel-set and bogie to reduce vibrations from wheel/rail interaction.
2nd stage		Secondary suspension: suspension system between the bogie and car body, aiming at reducing the vibrations of lower frequency.
3rd stage	Track	Rail pad and fastener
4th stage		Ballast/sub-ballast layers

As described in par. 13.1.3, the reduction of the pressure (due to the forces) under the ballast is about 3400-4200 times. According to Esveld [26]) there are some similarities in modelling of vehicle and track, as shown in the following table.

Table 9 Similarities in modelling of vehicle and track

	Vehicle	Track
Primary suspension	Suspension system between the wheel-set and bogie	Rail pad and fastener
Secondary suspension	Suspension system between the bogie and car body	Ballast / sub-ballast layers

15.3.1 Models of Track Dynamic Behaviour

Models of track dynamic behaviour may be generally classified into two categories:

- those that represent the track as a **continuously supported rail beam** and
- those that represent the track as a **discretely supported rail beam**.

Continuously supported models of infinite length are based on the beam on elastic foundation theory.

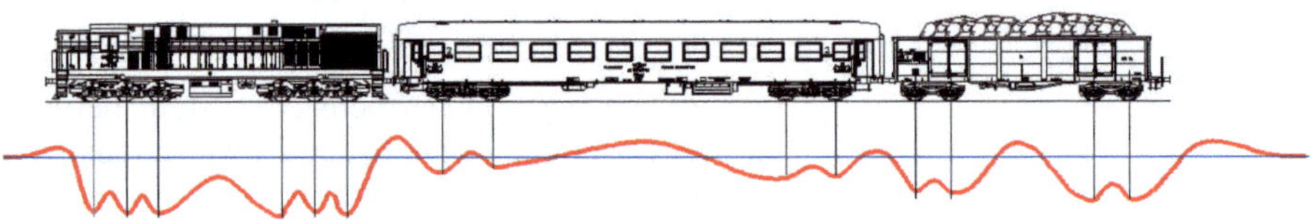

Fig. 32 The rail deflection due to a running train (after [12])

The rail, pads, fastenings, sleepers, ballast, sub-ballast and subgrade are components that define the value of the modulus of track elasticity (see Chapter 16).

Discretely supported models are similar to the continuously supported models and often have multiple layers representing the rail pads, sleepers, ballast, sub-ballast and subgrade.

According to this presentation, **a model** for **the entire railway system** can be developed consisting of

- **masses,**
- **springs and**
- **dampers** (Figure 33).

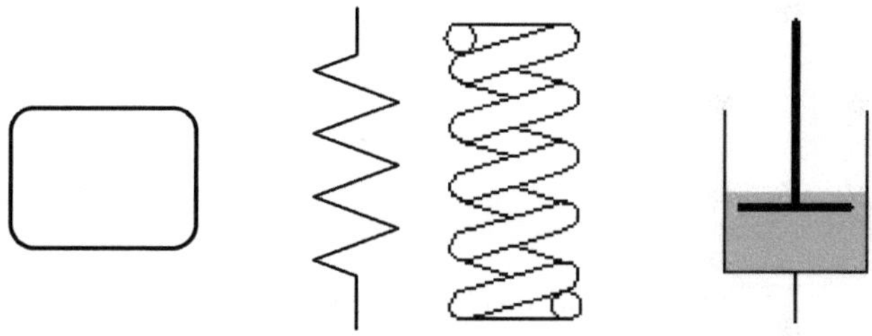

Fig. 33 Symbols used for the constituents of the model for the railway system

15.3.2 Continuously Supported Models: Rail as a Beam on Elastic Foundation (BOEF)

Rail as a Beam on Elastic Foundation may be modelled as

- an ordinary Euler – Bernoulli beam[10] (the classical beam theory) or
- a Rayleigh – Timoshenko beam. The Rayleigh – Timoshenko beam theory includes the rotatory inertia of the beam cross section and beam deformations due to the shear force. Also, a longitudinal (axial) force in the rail may be included in these models.

Winkler (1867) used for modelling of the railway system, Euler – Bernoulli beam theory: an infinitely long rail beam resting on a uniform elastic foundation (Beam on Elastic Foundation (BOEF) model).

The rail is modelled as a beam (with bending stiffness EI) which rests on a continuous elastic foundation (Figure 34). The elastic foundation represents all track components and is modelled by evenly distributed linear spring stiffness. The distributed force supporting the beam is then proportional to the beam deflection (Figure 35).

Based on this model, the deflection and bending moment of the rail beam due to the moving load can be determined.

[10] The Euler–Bernoulli beam theory (or classical beam theory or just beam theory) is a simplification of the linear theory of elasticity which provides a means of calculating the load-carrying and deflection characteristics of beams. It covers the case for small deflections of a beam which is subjected to lateral loads only. The Euler-Bernoulli equation describes the relationship between the beam's deflection and the applied load.

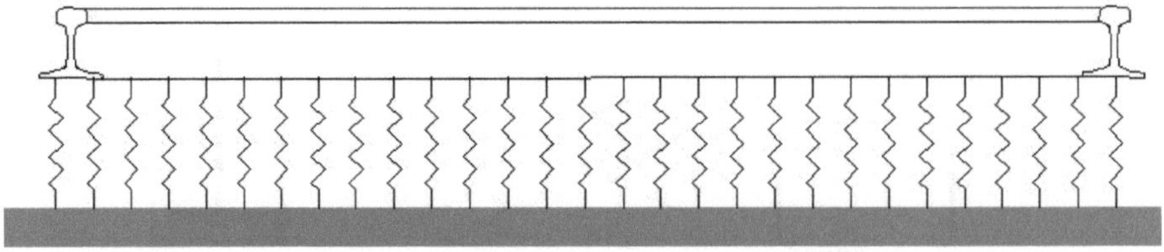

Fig. 34 Beam on elastic foundation (unloaded)

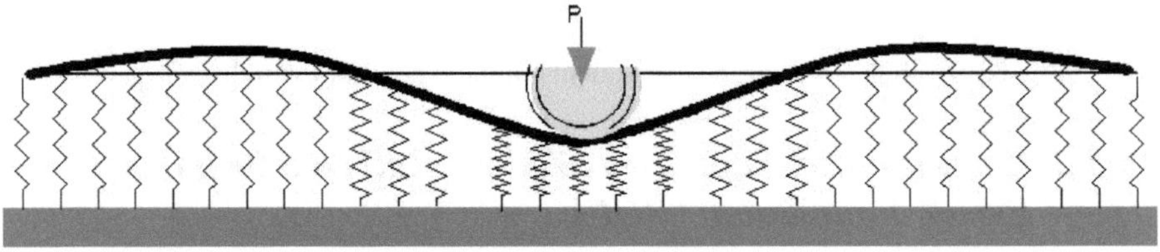

Fig. 35 Beam on elastic foundation loaded with a point force P from the wheel. The thick line indicates beam (rail) deflection due to the wheel load P.

This is the simplest track model and is still in use for easy and quick track deflection calculations for track design and analysis purposes.

The **vertical deflection** *y*, for a given point load *P* as a function of the distance from the load *x*, is:

$$\mathbf{y(x)} = -\frac{\mathbf{P \cdot \beta}}{\mathbf{2u}} \cdot \mathbf{e^{-\beta \cdot x}} \cdot [\mathbf{\cos(\beta x) + \sin(\beta x)}]$$

The bending moment is

$$\mathbf{M} = -\frac{\mathbf{P}}{\mathbf{4\beta}} \cdot \mathbf{e^{-\beta \cdot x}} \cdot [\mathbf{\cos(\beta x) - \sin(\beta x)}]$$

Where:

$$\boldsymbol{\beta} = \left(\frac{\mathbf{u}}{\mathbf{4EI_z}}\right)^{\frac{1}{4}}$$

and

- P is the load
- *x* is the distance along the track
- *E* is the Young's modulus of elasticity of the rail steel (beam) and
- I_z is the second moment of area of the beam.
- *u* is the estimate of the track modulus (see Chapter 16).

With this model it is also possible to calculate, for example, the individual sleeper reaction (by multiplying by the sleeper support stiffness and the sleeper deflection), due to the fact that load is proportional to the deflection. The load is shared mostly between three sleepers, with the sleeper beneath the load taking 40-50% of the axle load.

The BOEF method was originally developed for longitudinally sleepered tracks, and has since been applied to transversely sleepered track. The BOEF model is generally regarded as the most acceptable method for the analysis of rail foot stress and rail deflections.

This model may be used only for static loading of a track on a soft support (i.e. a track with wooden sleepers). **Dynamic effects cannot be analysed** using this model as it contains no mass.

Especially, the pinned – pinned frequency at approximately 800 to 1000 Hz, when the rail vibrates with nodes at the sleepers, is not contained in this model.

To simulate this model the function of rail pads, sleepers, and ballast, we have to introduce several continuous layers instead of the single layer shown in Figure 34 and Figure 35.

A continuous model like this one was used for the examination of the behaviour of the track support [71]. Effects of changes in the support (the rail pads) were studied, and it was found that a softer rail pad isolates the sleeper more effectively and it will significantly reduce the sleeper strains [32].

The classical representation of the track as a simple BOEF with a single moving load provided an initial understanding of track vibration at the steady-state. It is however an oversimplification of track modelling when subjected to high frequency impact associated with modern-day railway tracks. Several deficiencies arise from the BOEF model:

- Sleeper mass and bending flexibility are ignored;
- The shear distortion and rotary inertia of the rail, that are important factors for high frequency vibration, are not accounted by the simple Bernoulli- Euler theory;
- The discrete rail supports from individual sleepers are neglected and replaced by a uniform underlying foundation;
- Impact or impulsive loading associated with high frequency vibration is not considered; and
- Detailed dynamic behaviours of track components are not obtained.

The range and applicability of the one-dimensional theory of beams can be extended by taking account of transverse shear deformations and rotary inertia in the case of vibrating beams.

The limits of use the BOEF model are presented in the following table:

Table 10 Limits of use the BOEF model

Use of the model	Disadvantages of the model
Use only for static loading of a track on a soft support (i.e. a track with wooden sleepers)	Dynamic effects cannot be analysed
	The pinned – pinned frequency (when the rail vibrates with nodes at the sleepers) cannot be described

15.3.3 Discretely Supported Models

15.3.3.1 Overview

Modern railway tracks are characterised by accurately positioned continuous welded rails and sleepers. This makes the mechanical modelling of railway track rather simple: two straight parallel beams supported by sleepers at equal distances (Figure 36).

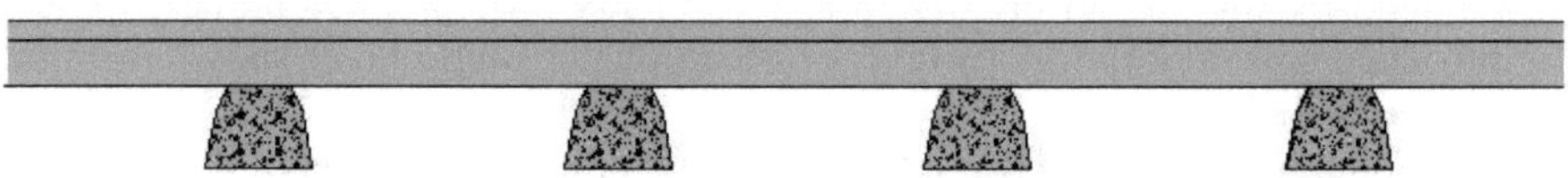

Fig. 36 Rail as straight parallel beam supported by sleepers at equal distances

Discretely Supported Models are similar to the continuously supported models but

- they consider the discrete spacing of sleepers.
- they often have multiple layers representing the Rail pads, sleepers, ballast, sub-ballast and subgrade.

As a consequence, the rails will obtain vibration modes related to this type of supporting. The most important vibration mode will be like a kind of bending between discrete points or pins, where the rail vibrates with nodes at the sleepers (Figure 37). With some simplifying assumptions, this pin-pin vibration resonance occurs at a specific frequency f_{pp}, which can be calculated by [33]:

$$f_{pp}=\frac{\pi}{2l^2}\sqrt{\frac{EI}{m}}$$

Where

I : distance between two supports [m]
EI: bending stiffness of the rail (static) [Nm^2]
m : mass of the rail per unit length [kg/m']

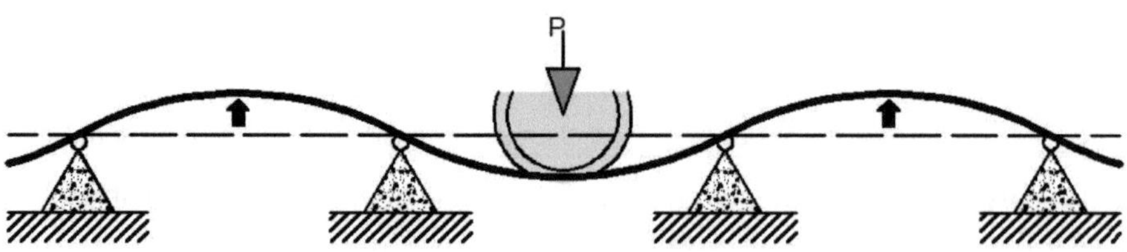

Fig. 37 Pin-Pin vibration mode

The equation is a general one, based upon an infinitely long beam and perfectly supported at small but rigid points.

This assumption doesn't correspond to reality: the rail sits elastically on the sleeper with an elastic rail pad in-between. Also the sleeper seats elastically in the ballast.

These and other elements may influence pin-pin resonance. These elements introduce new significant vertical resonances. For most track structures these three vertical resonances are dominant and lie between 30 and 2000 Hz.

15.3.3.2 Pin-Pin Resonance as a Reference in Determining Ballasted Railway Track Vibration Behaviour

Pin-pin resonance is one of the most significant preferred vibration modes of beams, which are supported at equal distances, such as rails at sleepers in railway track structures do [32].

To include also the pinned – pinned frequency in the track model, the continuous rail should be supported at discrete points.

The supports could then be either discrete spring – damper systems or spring – mass – spring systems, modelling Rail pads, sleepers and ballast bed.

One commonly used method to model this is to place the rail (a beam) on a spring and a damper in parallel: by this **spring – damper system** (see Figure 38):

- rail is modelled as a beam
- the rail pads are modelled by spring – damper systems,
- the sleepers are rigid masses, and
- the ballast -as an elastic foundation- is modelled by spring – damper systems

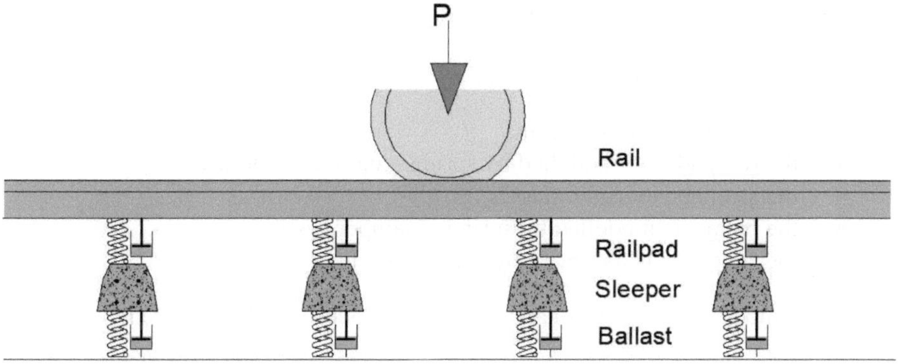

Fig. 38 Rail on discrete supports

Using this track model, the three resonance frequencies can be captured, namely

- the track structure bouncing on the ballast,
- the rail and sleeper vibration with the rail pad as a spring between the two masses, and
- the pinned – pinned frequency.

15.3.3.3 Discretely Supported Track Including Ballast and Subgrade

The ballast and subgrade masses influence each other, indicating that a deflection at one point (at one sleeper) will influence the deflection at the neighbouring sleepers.

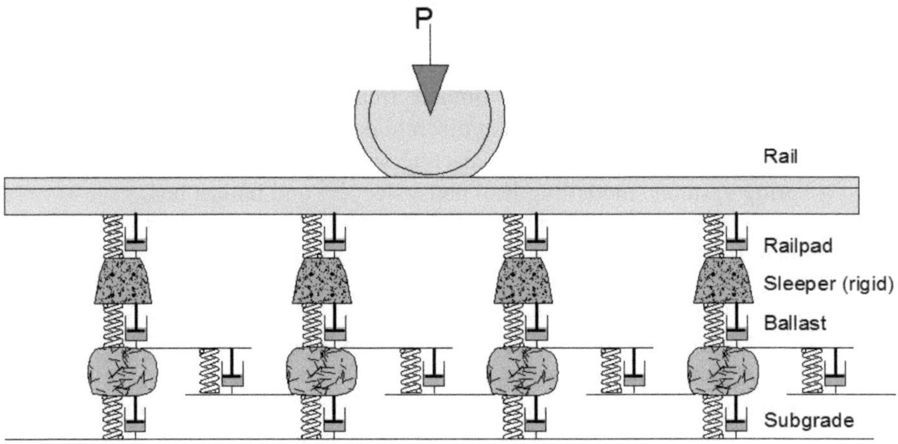

Fig. 39 Rail on discrete supports with rigid masses modelling the sleepers [32]

By this **spring – damper system** (see Figure 39):

- rail is modelled as a beam,
- the rail pads are modelled by spring – damper systems,
- the sleepers are rigid masses,
- the ballast is modelled by spring – damper systems
- the subgrade is modelled by spring – damper systems

By this model, the four resonance vibration modes may be captured:

- embankment vibration,
- track-on-the-ballast vibration,
- rail-on-rail pad vibration, and
- pinned – pinned vibration of the rail

15.3.4 Vehicle – Bridge Interaction (Train Passing a Bridge)

When a train enters a bridge, the change of the stiffness on the transition zone between the plain track and the bridge (on plain track i.e. $E \approx 25*10^6$, on the bridge $E \approx 100*10^6$) often causes problems. This change of stiffness causes increased dynamic forces, which depends on the speed, the stiffness ratio, the damping properties and the transition length (Figure 40) [46].

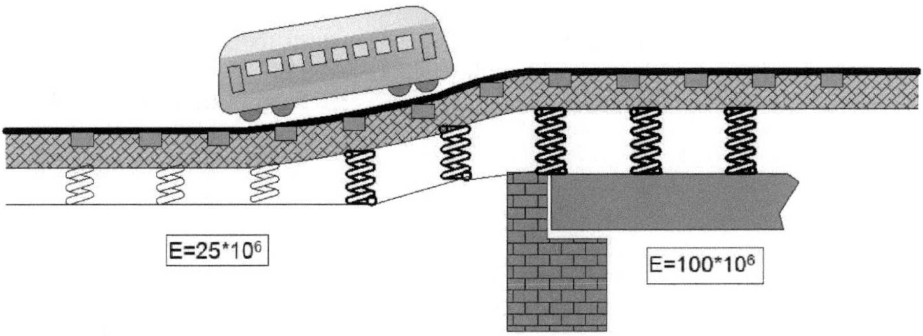

Fig. 40 A train passing a bridge

Not only a track, but also a railway bridge is a dynamic system. Both systems interact when a train is passing the bridge. This can be modelled as moving mass on simply supported beam (Figure 41).

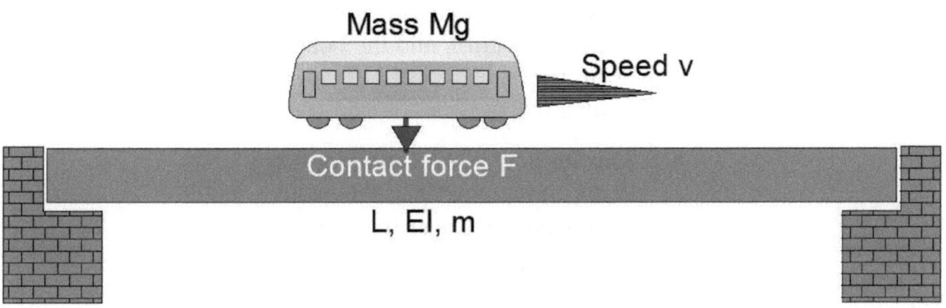

Fig. 41 Contact force between a moving mass M and a simply supported beam [32]

The overall behaviour of the beam – mass system when a train is passing the bridge will be described below:

Table 11 The overall behaviour of the beam – mass system when a train is passing the bridge (after [9])

Event	What happens	Why it happens
The wheel enters the bridge	the contact force equals the static load of the wheel i.e., $F(0) = M_g$ ⇩	
	immediately when the wheel starts to load the bridge, the contact force F decreases ⇩	the bridge (the beam) starts to deflect under the horizontally moving mass (the wheel)
	• the dead load M_g of the mass becomes larger than the reaction force F(x) and • the mass will start to move downwards (it follows the deflected beam) ⇩	because of the decrease of the contact force
	After a while the beam deflection will change direction and the beam moves upwards, also pushing the mass upwards: • This induces a large contact force between the mass and the beam • F(x) becomes larger than M_g	
The wheel reaches the position x= 0.82 L	The wheel – beam contact force has its maximum, which is almost 2.5 times the static load M_g. This causes the wheel to continue moving upwards, and, the wheel will lose contact with the beam just before it reaches the support to the right.	

A further discussion of the transition of track and bridge is presented in par. 16.7.

16 The Effect of Track Stiffness on Track Performance

> *Key terms:*
>
> *track stiffness, track modulus, vertical deflection, rail deflection, global track stiffness, deterioration process, transitions on the railway track*

16.1 Track Stiffness – Track Modulus (An Overview)

What is track stiffness? There is not a complete consensus on definitions of track stiffness. The most general understanding defines track stiffness as the elastic rail deflection that takes place under a wheel loading.

This generic track stiffness is a function of the structural properties of the

- rail,
- resilient pads in the track structure,
- sleepers,
- ballast,
- sub-ballast and
- subgrade soil(s).

Fundamental analysis of track stiffness has typically been approached by means of the Beam on Elastic Foundation (BOEF) approach (see par. 15.3.2).

This analysis technique is known to be reasonably representative, but introduces the concept of the track modulus, i.e. the stiffness of the spring k per unit length of track.

Track stiffness	Track modulus
Track stiffness refers to the rails. It is • the elastic rail deflection that takes place under a wheel loading. • the ratio of the load applied to the rail to vertical rail deflection	Track modulus refers to the track. It is • the stiffness of the spring k per unit length of track. • a measure of the vertical stiffness of the track foundation

This constant k is that which would be obtained by applying a uniformly distributed load along the whole track and recording the deflection. The modulus

K. Tzanakakis: *The Railway Track and Its Long Term Behaviour*, STTT 2, pp. 79–87.
DOI: 10.1007/978-3-642-36051-0_16

therefore represents the equivalent series stiffness of all the components of the track.

It is:

$$\frac{1}{k_{total}} = \sum_{i=1}^{n} \frac{1}{k_i} = \frac{1}{k_{rail}} + \frac{1}{k_{pad}} + \frac{1}{k_{sleeper}} + \frac{1}{k_{ballast}} + \frac{1}{k_{substructure}}$$

where k_i is the coefficient of "Rail Support Modulus"[11] of each layer. This implies that k_{total} is a coefficient of quasi elasticity (stiffness) of the track, the equivalent of the "spring constant" in Hooke's law. It is defined as the "reaction coefficient of the sleeper", and ρ_i is the "spring constant" of each layer [134].

Track stiffness includes the bending stiffness of the rail, whereas track modulus is concerned only with the support condition below the rail.

Both track modulus and track stiffness are related to the track performance.

16.2 The Calculation of Track Modulus

The most commonly known method for the calculation of track modulus is the beam-on-elastic foundation method (par. [15.3.2]). A vertical force P (wheel) produces a vertical rail deflection y.

The track stiffness *k*, taken at a point as the wheel passes directly overhead, is defined as

$$k = \frac{P}{y}$$

The **vertical deflection** *y*, for a given point load *P* as a function of the distance from the load *x*, is (see also par. 15.3.2):

$$\mathbf{y(x)} = -\frac{\mathbf{P \cdot \beta}}{\mathbf{2u}} \cdot \mathbf{e}^{-\boldsymbol{\beta} \cdot \mathbf{x}} \cdot [\mathbf{cos(\beta x)} + \mathbf{sin(\beta x)}] \qquad [42]$$

Where:

$$\boldsymbol{\beta} = \left(\frac{\mathbf{u}}{\mathbf{4EI_z}}\right)^{\frac{1}{4}} \qquad [42]$$

and

- *E* is the modulus of elasticity of the rail steel (beam) and
- I_z is the second moment of area of the beam.
- *u* is the estimate of the track modulus.

Figure 42 shows the results of this model. The deflection of the rail relative to the ground is shown for a fully loaded by a freight wagon for various track moduli. The maximum absolute deflection occurs at the wheel/rail contact points for high modulus (stiff) track and at the midpoint between the wheels for low modulus (soft) track.

---

[11] c in German literature, k in American literature, ρ In French literature.

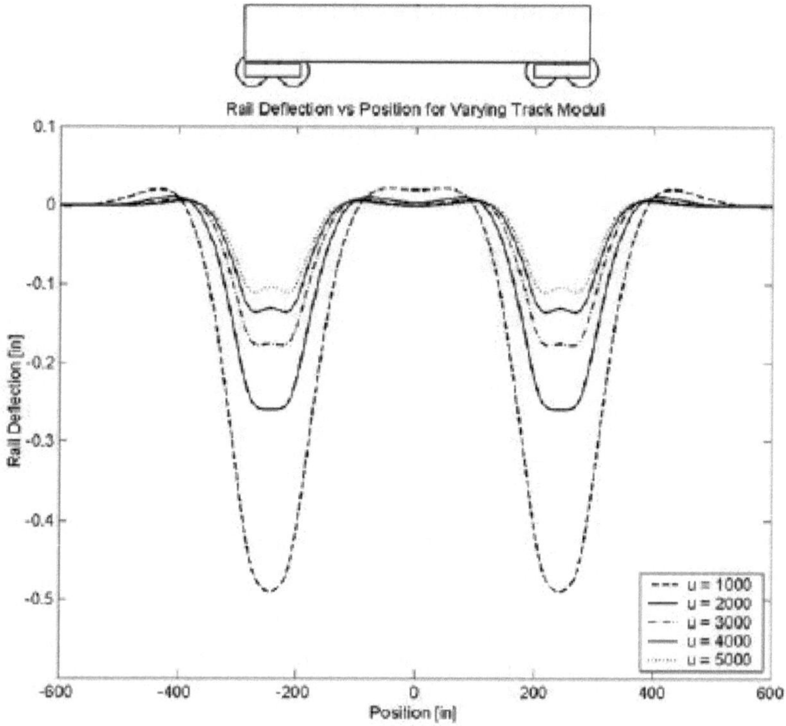

Fig. 42 Winkler Model of Rail Deflection ([42])

Typical track stiffness values are shown in Table 12.

Table 12 Typical Track Stiffness Values (after [50])

	Stiffness [MN/m]
Soft Track bed	10
Typical Track bed	40
Stiff Track bed	100
Soft Rail pad	75
Typical Rail pad	150
Stiff Rail pad	500
All values per sleeper end	

16.3 Track Stiffness and Deterioration

Global track stiffness is a combination of stiffnesses of all different layers and components.

The way local stiffnesses are combined determines the displacements of each layer. Forces and displacements will together with support conditions determine bending moments, as described in the following cases:

Case	Influence / Result	Explanation
Track of low global stiffness	• Large displacements and large bending moments of the rail. • Dynamic train-track interaction forces will be relatively low. • Influence on long-term fatigue conditions.	Large rail displacements: more sleepers will be involved sharing the load leading to lower forces on sleepers.
Track of high global stiffness	• Small displacement and small bending moments of the rail. • Dynamic train-track interaction forces will be relatively higher.	Small rail displacements have the effect that fewer sleepers will share the load increasing forces for each sleeper.

Ballast and sub-layers are influenced both by stiffnesses above and under themselves. A ballast layer resting on bedrock will experience large internal compression/expansion during a train-passage as compared to a ballast layer resting on a soft clay layer.

A differential settlement in the ballast layer is more connected with compaction and variations in sleeper support. Two soil layers can have the same stiffness, but totally different ability for settlements. Yet, very soft layers, such as clay and peat, have generally close correlation between stiffness and settlements.

The track stiffness experienced by a train will vary along the track. Sometimes the stiffness variation may be very large within a short distance [48].

Examples:

(1) Unsupported sleeper: Track stiffness is locally at that sleeper, very low.
(2) Insulated joints: the bending stiffness of the rail has a discontinuity implying a discontinuity also of the track stiffness.
(3) Transition from an embankment to a bridge (see par. 15.3.4).
(4) At switches both mass and stiffness change rapidly (see par. 17.6.3.3).

Changes of track stiffness will induce changes in the wheel/rail contact force. This will intensify track degradation such as increased wear, fatigue, track settlement due to permanent deformation of the ballast and the substructure, and so on.

As soon as the track geometry starts to deteriorate, the changes of the wheel/rail interaction forces will increase, and the track deterioration rate increases (Figure 43).

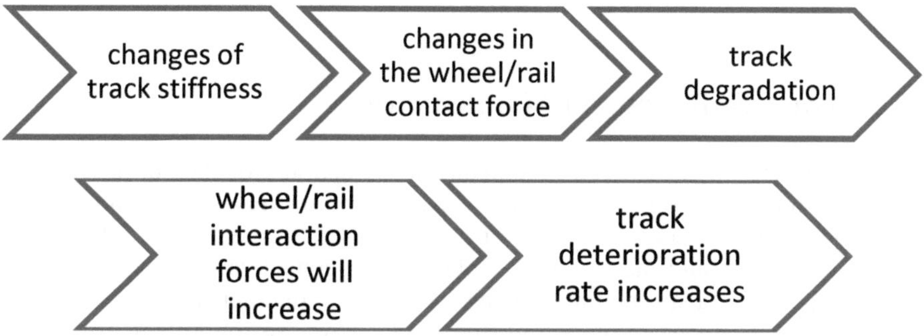

Fig. 43 The deterioration process due to variation in track stiffness

The rate of degradation of track components and the rate of track settlement will depend on the severity of the stiffness variation. As soon as the track geometry starts to deteriorate, the variations of the train/track interaction forces increase, and this speeds up the track degradation rate.

Therefore, the influence of track stiffness irregularities on the development of track settlement and on the deterioration of track components and materials cannot be ignored.

By modifying the stiffness variations along the track, for example by use of grouting or under-sleeper pads, the variations of the wheel/rail contact force may be considerably reduced.

16.4 Track Stiffness Irregularities and Its Origin

Track stiffness irregularities may have its origin in the

- track superstructure (rails, rail pads, sleeper, ballast) or
- substructure (foundation, subgrade soil, etc.).

An example for cause an irregular stiffness of the substructure and of the ballast could be a non-uniformly compacted ballast lying on substructure, with properties varying along the track. In this case, the track stiffness experienced by a train will also vary along the track.

Places along the track where track stiffness will change rapidly are for example at

- pile decks,
- embankments, bridges (the transition area from an embankment to a bridge is a place where severe track settlement may occur) (see par. 15.3.4)
- transition zones etc. (see par. 15.3.4 and par. 16.7)

- switches and turnouts, especially at the crossings (the frogs) (see par. 17.6.3.3),
- insulation joints,
- hanging sleepers (the track stiffness changes very rapidly) (see par. 19.7.2).

16.5 Track Stiffness and Maintenance

In order to determine the necessary maintenance actions for the substructure, its condition must be known. Usually most attention has been paid by the Infrastructure Managers to inspection techniques of the superstructure. Especially for the sub-ballast and the subgrade (subsoil) less consideration has been given, although they have a substantial influence on the track maintenance costs.

The vertical stiffness of a railway track plays an important role when considering maintenance work, but it is also an important factor when looking at dissipated energy of a train.

Optimal values of the vertical stiffness are proposed in [63]. Optimal values optimize maintenance costs and the costs for dissipated energy of a train versus the vertical track stiffness. According to [63] the optimum vertical track stiffness should be between 70 and 80 kN/mm.

UIC Leaflet 719R [12] gives advice on the construction of earthworks for railway lines and considers also the special case of transitions to structures.

> Railway track stiffness (vertical track load divided by track deflection) is a basic parameter of track design which influences
>
> - the bearing capacity,
> - the dynamic behaviour of passing vehicles and, in particular,
> - track geometry quality and the life of track components.

16.6 Smoothing Track Stiffness Irregularities

In order to construct a transition zone with the desired variation of stiffness and geometry there are some possibilities:

- Use of elastomeric products, such as rail pads, under sleeper pads (USP), and sub-ballast mats (SBM),
- Use of geogrid (or geotextile) reinforcements
- Chemical treatment of sub-soil
- Management of the ground water
- Use of fastenings with varying elasticity

[12] *UIC Code 719R* "Earthworks and Trackbed Construction for Railway Lines".

16.7 Changes of Track Stiffness in Transitions on the Railway Track

16.7.1 Forces at the Transition Zones

Transition zones are zones in which the track stiffness changes abruptly. Transition zones are usually associated with track geometry deterioration that requires frequent maintenance. This change of stiffness causes increased dynamic forces, which depends on the speed, the stiffness ratio, the damping properties and the transition length.

Transition zones can usually be found near bridges, abutments and tunnels. As discussed in par. 15.3.4, when a train enters a bridge, the change of the stiffness on the transition zone between the plain track and the bridge (on plain track i.e. $E \approx 25 \text{ X } 10^6$, on the bridge $E \approx 100 \text{ X } 10^6$) often causes problems. The observed problems in the transition zone can include [49]

- ballast penetration into the subgrade,
- hanging sleepers (i.e. voids forming underneath the sleepers) (see 19.7.2),
- permanent rail deformation
- worn track components
- cracking of the sleepers (and/or slab track) and
- loss of gauge

16.7.2 Railway Track Transition Mechanics

Railway track transition mechanics can be explained as follows ([49], see also par. 15.3.4):

(1) At transition points the trains wheel will experience rapid changes in elevation (either up or down) due to abrupt changes in track stiffness
(2) The vehicle components start to excite (wheel, boogies and car body)
(3) The excitation of the track components leads to vertical vehicle – track interaction forces, depending on

   a. the elevation change, which is related to the change of the stiffness
   b. the train speed
   c. the vehicle mass
   d. the suspension characteristics of the vehicle

In the transition zone, the track may settle more than on the plain line track (before entering the bridge or the tunnel) than at the rigid base (bridge or tunnel), where the track is stiffer. This can lead to dips in the track position (running surface) and hanging sleepers (see 19.7.2). In cases of soft subgrade soils and / or poor drainage, the settlement close to the transition zone can be significantly accelerated.

Due to the relationship between the dynamic load of the running trains and the track deflection, the track geometry **in the transition zone can deteriorate at an accelerating rate through a self-perpetuating mechanism: as the dynamic load increases so does the track differential settlement, which in turns leads to even higher dynamic forces and so on** [49].

16.8 The Role of the Track Components for the Track Stiffness

The track components, as part of the track-structure, determine the track stiffness. The track stiffness derives from the series value of the individual components [50]:

Rail	The relationship between track modulus and track stiffness is determined by the rail bending stiffness.
Rail pad	There is a greater tendency towards softer Rail pads, particularly for high speed application, as they obviously reduce dynamic impact forces.
Baseplate pads 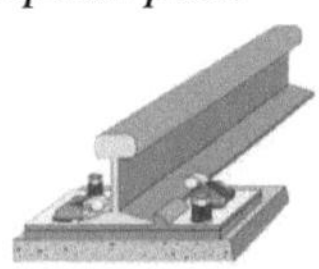	Baseplate pads provide greater track flexibility without excessively compromising rail fastening arrangements, primarily for the reduction of ground vibrations.
Sleepers	Concrete sleepers are basically rigid. With timber sleepers the effective track stiffness is certainly lower.
Sleeper Spacing	Reducing intervals between sleepers increases track stiffness.
Under sleeper pads	They are used to reduce ground vibration or reduce track stiffness on ballasted bridges.
Ballast	As a bulk material ballast is relatively stiff and its stiffness depends to a certain extent on the degree of compaction.

16.9 Influences on Track Stiffness

16.9.1 Direct Influences of Track Stiffness

Parameters affecting direct track stiffness are:

- A stiff rail distributes the load over a greater number of sleepers, whereas a less stiff track modulus distributes the load over fewer sleepers (this would mean higher sleeper and rail fastening forces).
- A stiff rail results in lower rail deflection and lower (static) bending stresses in the rail.
- A stiff track modulus results in lower (static) rail deflection and stresses.

- Low Track bed stiffness directly produces an increase in ballast settlement [68].

16.9.2 Long-Term Influences of Track Stiffness

The three primary areas of influence of the track stiffness on track life-cycle behaviour seem to be [50]:

- Track geometry deterioration
- Rail fatigue
- Other component deterioration

For track geometry, the track bed (and particularly the sub-soil) stiffness appears to be most relevant. Differential track settlement is the primary cause of track geometry deterioration. The track stiffness can vary due to a variety of reasons such as:

- Geology
- Moisture content
- Embankment material
- Ballast and sub-soil layer depth
- Structures

16.10 Track Stiffness and the Influence on Noise and Vibration

Track stiffness requirements for noise and vibration may differ from those for track life cycle performance [50].

For noise the stiffer the rail pads and the softer the ballast produces the lowest noise levels. However, the surface roughness of the railhead is significant for the amount of noise radiated by the wheel and rail. Thus, any changes to the stiffness of the Rail pad and the ballast that increase the growth of rail head roughness is, in the long term, likely to be important.

With ground borne noise, ground borne vibration and bridge noise, the response of the ground and the building are significant. This means that the optimum stiffness for the rail pads and ballast will be specific to a site. However, the overall condition of the track has a significant influence in the amount of vibration and noise entering the ground. Therefore, in the long term any influences that rail pad stiffness and ballast stiffness have on the track condition are likely to be significant [50].

17 The Wheel-Rail Interface

Key terms:

wheel/rail contact, wheel tread, wheel flange, contact area, normal contact, Hertzian contact, Hertzian Contact Theory, Hertzian spring, tribology, friction, lubrication, wear, short wavelength irregularities, long wavelength irregularities, impact loads, oscillations, vibrations, and noise, wheel out-of-roundness, rail bending stiffness, wheel flats, rail joints, wheel wear, friction control, friction modifiers, contact stresses, Active control of vehicle dynamics, wear-resistant wheel profiles, rail grinding

17.1 Overview

The wheel/rail contact determines the capabilities of the railway system. This contact is the critical point at high speeds and in cases of heavy freight trains with high axle load, due to the high contact forces and the dynamics, but it is also responsible for noise emissions and wear (Figure 44).

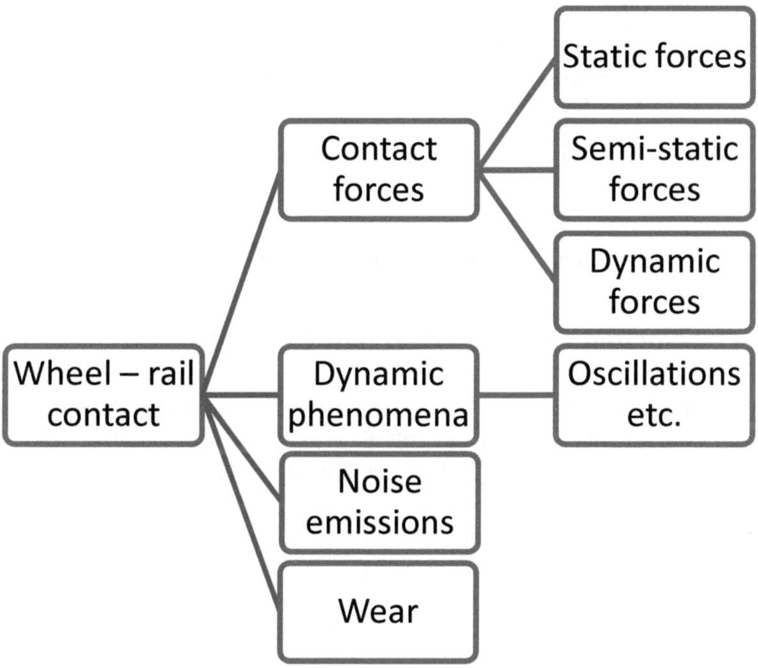

Fig. 44 Wheel-rail contact impacts

K. Tzanakakis: *The Railway Track and Its Long Term Behaviour*, STTT 2, pp. 89–112.
DOI: 10.1007/978-3-642-36051-0_17

Figure 45 shows the parts of a wheel at the rail interface.

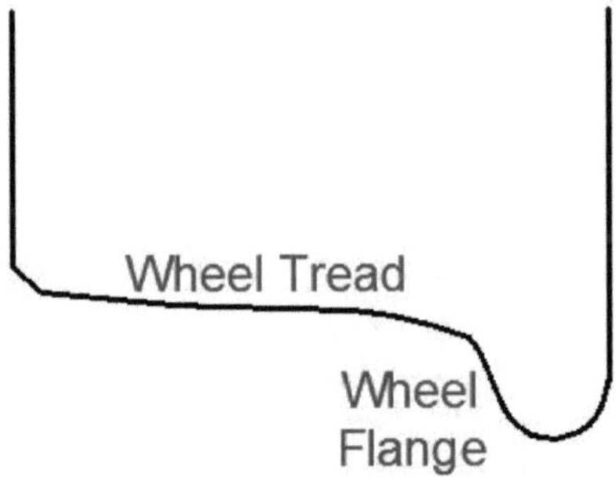

Fig. 45 Parts of a wheel at the rail interface

As we mentioned in 13.1.2, the outer rail in a curve is situated higher in a curve due to canting and it is called the "high rail". The inner rail is lower than the outer rail and is therefore called the "low rail".

The inside of the rail head facing the other rail is called the "gauge face" and the outside of the rail head is, for obvious reasons, called the "field face" (Figure 46).

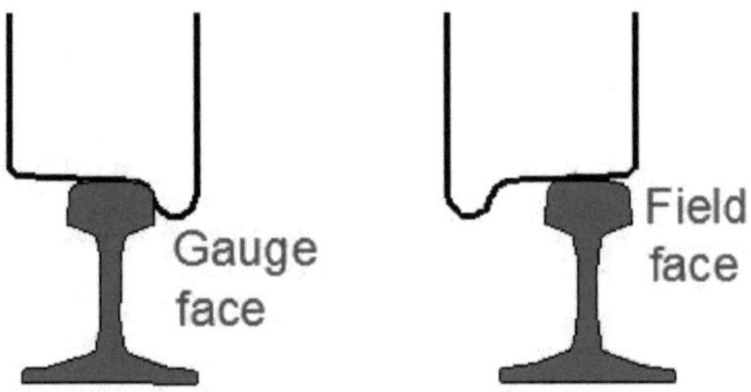

Fig. 46 Gauge face - Field face of the rail

17.2 The Normal Contact (Hertzian Contact)

The study of the contact between bodies is possible today with finite element methods. However, the necessity to calculate as quickly as possible leads to the use of analytical methods [32]. This section describes the classical Hertzian model.

Railway track dynamic models typically use the Hertzian Contact Theory to represent the wheel-rail contact (Figure 47).

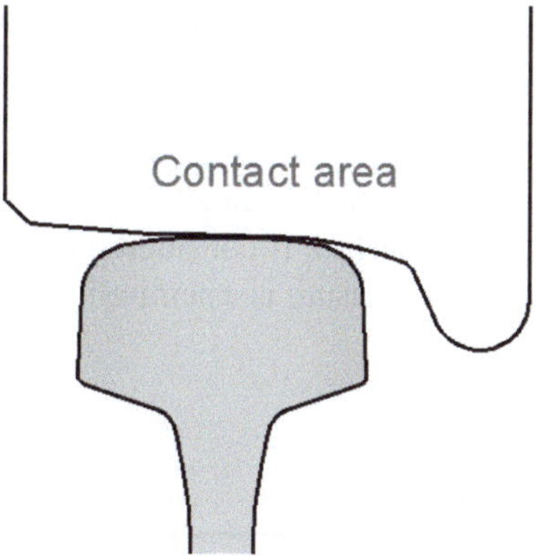

Fig. 47 The wheel - rail contact

Hertzian contact stress refers to the localized stresses that develop as two curved surfaces come in contact and deform slightly under the imposed loads (Figure 48). This amount of deformation is dependent on the modulus of elasticity of the material in contact. It gives the contact stress as a function of the normal contact force, the radii of curvature of both bodies and the modulus of elasticity of both bodies.

Fig. 48 Stresses in a contact area loaded simultaneously with a normal and a tangential force. Stresses were made visible using photo elasticity

Hertz[13] demonstrates that ([32])

when two elastic bodies are pressed together in the following conditions:

- Elastic behaviour
- Semi-infinite spaces

[13] H. Hertz, Über die Berührung fester elastischer Körper (On the contact of rigid elastic solids). In: *Miscellaneous Papers*. Jones and Schott, Editors, *J. reine und angewandte Mathematik* 92, Macmillan, London (1896), p. 156 English translation: Hertz, H.

- Large curvature radius compared to the contact size
- Constant curvatures inside the contact patch.

then:

- The contact surface is an ellipse
- The contact surface is considered flat
- The contact pressure is a semi-ellipsoid.

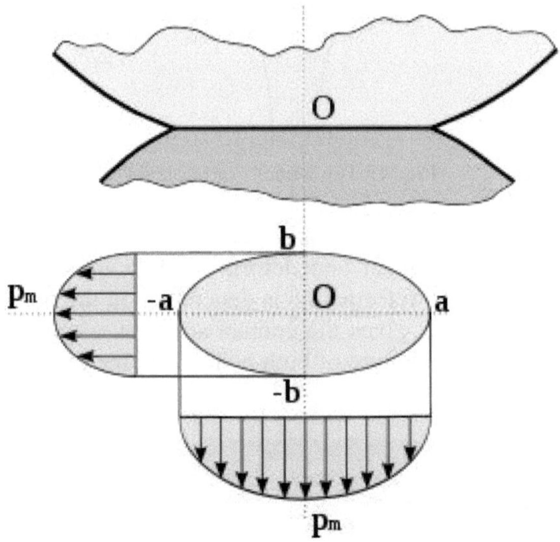

Fig. 49 Contact between two spheres

17.3 Train Running on Track

Vehicle and track can be seen as oscillatory systems. When a train is running, there are dynamic interactions between the two systems.

This dynamic interaction leads to forces between both systems and their components. Due to the physical damage caused by long-term process, changes are produced in the individual system characteristics, and thus lead to changes of dynamic loads.

The type of the interactions depends on the relevant system properties (such as mass, stiffness and damping, see Chapter 14) but also on the maintenance efficiency of the vehicles and the track.

When a train is running on the track, the springs and masses of the train[14] and the track[15] contribute to the dynamic behaviour of the whole system (train – track) and other resonance frequencies will appear (Figure 50). Dynamic effects in the complex train – track system become more significant when train speed increases and when the axle load gets higher.

---

[14] Wheel sets, suspensions, bogie frames.

[15] Rails, fastening system, ballast, subgrade.

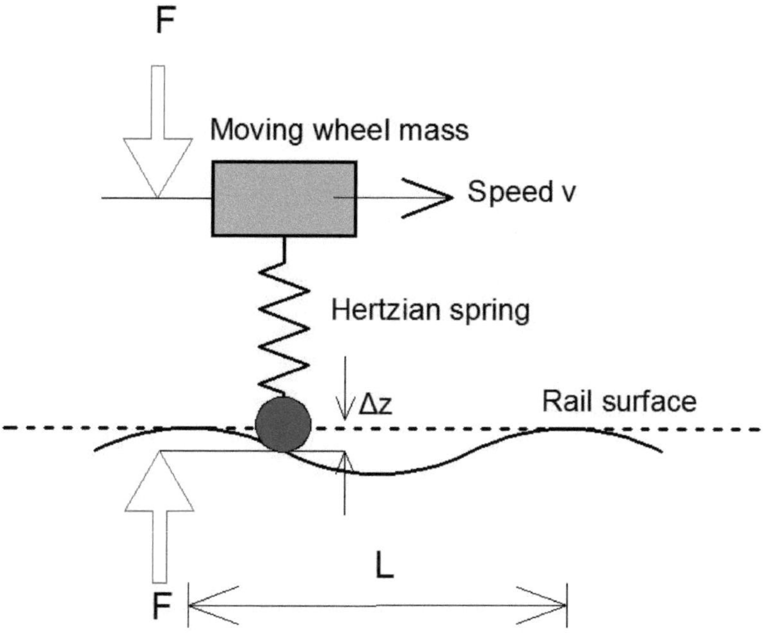

Fig. 50 Hertzian spring between wheel and rail

One "spring" is the "contact spring", at the wheel – rail contact (see Chapter 14 and Figure 19). The elastic deformation at this contact area can be seen as deformation of a spring with nonlinear characteristics. The stiffness of the "contact spring" increases with increasing wheel – rail contact force. This spring will influence the high-frequency behaviour of the wheel – rail system[16].

In Figure 51 an ideal situation is described: vehicle and track in a perfect condition. The track is loaded only by the static load, no dynamic effects influence the track.

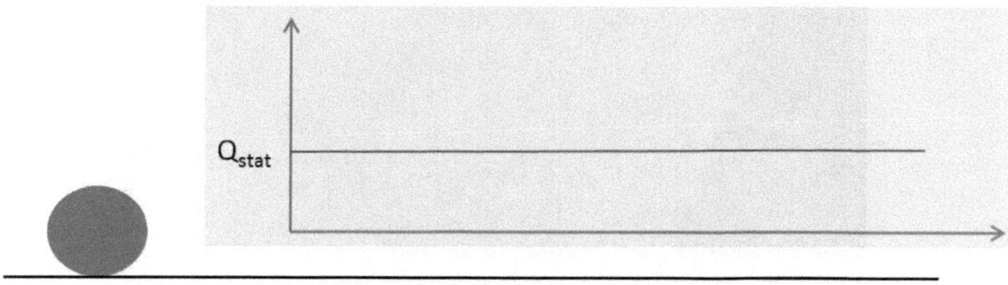

Fig. 51 Wheel and track in a perfect condition

[16] The spring is called a Hertzian spring, named after Hertz who investigated the contact between elastic bodies.

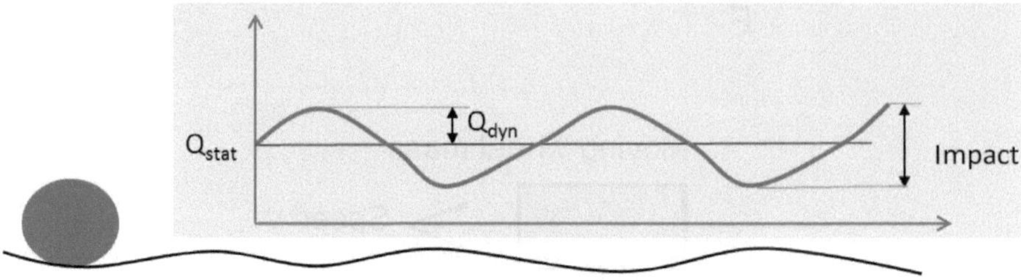

Fig. 52 Track with irregularities

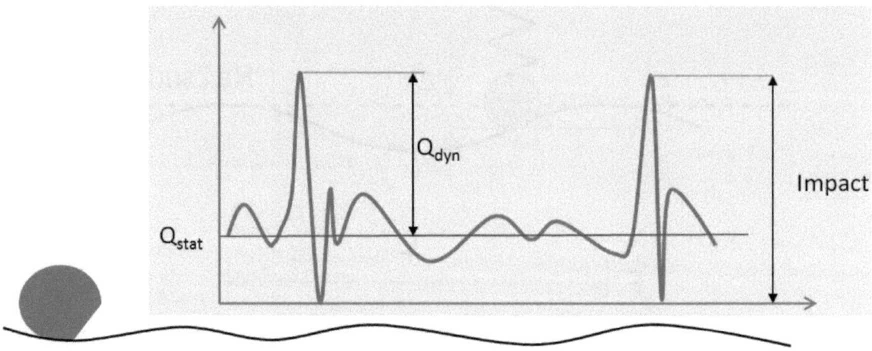

Fig. 53 Track and wheel with irregularities

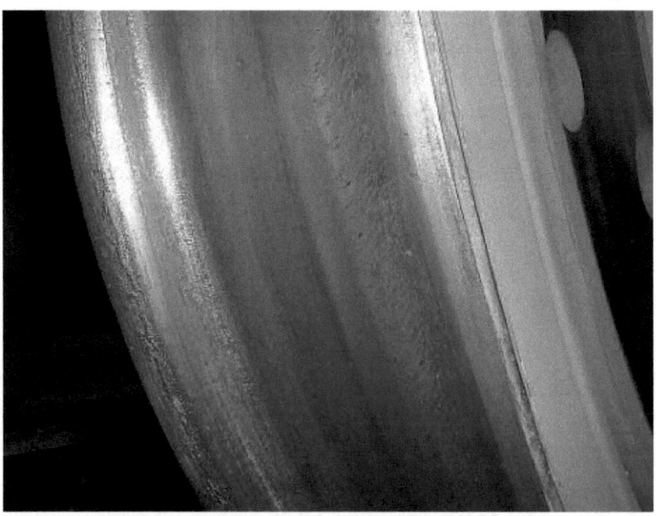

Fig. 54 Irregularities of track affect the wheel and wheel out-of-roundness affects the track quality

This situation is only theoretically possible. In reality we have track irregularities, which affects dynamic loads of the running train, additionally to the static load (Figure 52). This situation assumes that all the vehicles of a moving train have wheels without any irregularity, which do not correspond to reality. The wheel out-of-roundness causes additional dynamic forces, depending on the size of the wheel irregularities (Figures 53 and 54).

The following example demonstrates the impact of a track fault (a faulty weld on the rail) with changes in the individual system characteristics that lead to changes of the dynamic loads.

Example: a faulty weld on the rail:

Cause	*Impact*	*Explanation*
Faulty weld on the rail	*Forces at a running train*	*A faulty weld on the rail, induces forces at a running train*
	Wear on individual ballast stones	*The forces are transmitted through the sleepers to the ballast. In the ballast bed, as a result of the big forces, an increased wear on individual ballast stones arise*
	Track settlements	*The position of the stones with each other changes, which results in track settlements.*
	New interactions between the vehicle and the track	*Caused by the changed track position and changed ballast properties there are new interactions between the vehicle and the track*

Irregularities of track affect the wheel and wheel out-of-roundness affects the track quality.

17.4 Tribology

Tribology[17] is the science and technology of ([52])

- friction,
- lubrication and
- wear

In any situation where one surface rolls or slides over another, whether lubricated or not, it is affected by complex tribological interactions.

There are many parameters influencing the performance of a tribological system, either dry or lubricated.

[17] From the Greek "tribo" (τρίβω -meaning "I rub").

17.4.1 Friction

The friction force is the resulting resistance when a solid object in contact with another is moved tangentially with respect to the surface of the other [52].

The ratio of the friction force (F) to the normal load (P) is called the coefficient of friction (μ).

$$\mu = \frac{F}{P}$$

It is often desirable to have

- low friction, to save energy (high friction results in energy losses), or
- high friction, as in the case of brakes or in the running band of a wheel/rail contact.

Depending on the contact conditions, the friction can be divided into two main categories:

- rolling friction and
- sliding friction.

Within each main category there are further classifications depending on whether the interface between the surfaces is dry or lubricated. To be able to determine the friction coefficient for two surfaces in contact, a large number of variables need to be known. Due to the difficulties of determining all parameters, it is usually only possible to give rough estimation for the friction coefficient.

17.4.2 Wear

Wear is the process leading to loss of material from one or more solid surfaces in a sliding, rolling or impact motion relative to one another [52]. The wear causes mainly roughness by rail and wheel and changes also the respective profiles, which evokes new contact situations. Moreover, we have the fact, that railway vehicles especially in a narrow curve tend more to slip than to roll.

Theoretical prediction of wear is difficult and wear coefficients are often calculated from experiments to get accurate results.

17.5 Irregularities and Impact Loads Generating Excitation Frequencies

A wheel running at a speed v (m/s) over a sinusoidal rail irregularity of wavelength λ (m) will observe an excitation frequency f (in Hertz, Hz) that is:

$$f = \frac{v}{\lambda}$$

With vehicle speeds v = 5 to 60 m/s (20 to 220 km/h) and with irregularity wavelengths λ = 0,030 to 0,300 m (3 to 30 cm) one obtains excitation frequencies in the frequency interval f = 17 to 2000 Hz. Longer wavelengths give lower frequencies.

The excitation will then induce vibrations and noise in the train and in the track structure and the environment. Resonances in the train – track system will also amplify these vibrations at the resonance frequencies.

The sources generating train and track vibrations are:

- long wavelength irregularities and (Figure 55)
- short wavelength irregularities (railhead corrugation) (Figure 56),
- impact loads.

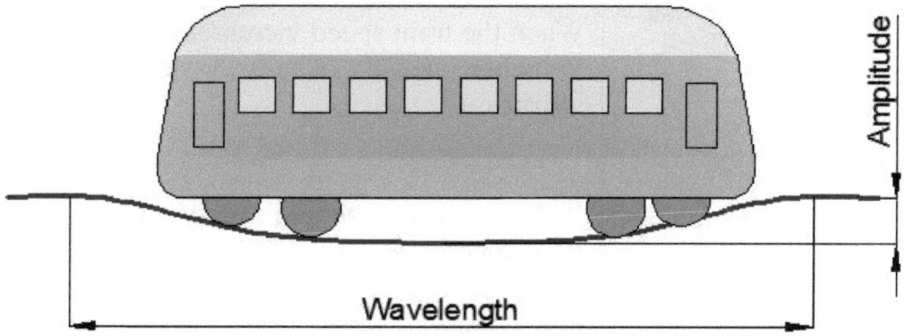

Fig. 55 Long wave track irregularities tends to cause body vibration, affecting the riding comfort

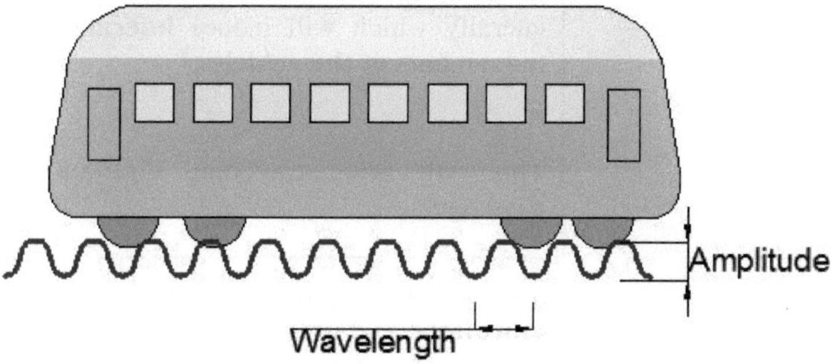

Fig. 56 Short wave length track irregularity tends to cause wheel vibration, affecting axle weight fluctuation and vibration

Especially at high speed, other dynamic phenomena appear. A large number of excitation sources exist that may induce **oscillations, vibrations,** and **noise** in the train and in the track and its surroundings (impact to passengers), as described in the following table:

Table 13 Causes of oscillations, vibration and noise and the impact to passengers (after [52])

Phenomena	Causes / Impacts for the passengers
Ground Vibrations and noise	Trains[18] induce **ground vibrations** that are transmitted to buildings along the railway lines. Noise and vibrations generated in the buildings, brings discomfort to people in them.
	When the train speed increases, it results generally to higher intensity of railway-generated **noise and vibration**. If (for example) the track support is soft[19], then the train speed may exceed the velocity of the waves in the ground[20]. This will bring a great increase of the vibration level, and train speed must be limited[21].
Long wavelength irregularities	Track settlement will induce long wavelength irregularities of the track both vertically and laterally which will induce **lateral and vertical movements of the vehicles**. This will bring travelling discomfort for the passengers.
Short wavelength irregularities	Short wavelength irregularities will induce **vibrations and noise**. This will bring travelling discomfort for the passengers.

[18] Mainly heavier and faster trains.
[19] The track is built on a layer of clay.
[20] Waves with both longitudinal and transverse motion: Rayleigh surface waves.
[21] This happened on the Swedish West Cost Line, where the track had to be removed and the subground strengthened before the trains were allowed to pass at high speed [32].

Irregularities in the track gauge	Irregularities in the track gauge will induce **lateral displacements** of the railway vehicles and influence the horizontal alignment of the track. This will bring travelling discomfort for the passengers.
Deficiencies or excesses in the cant	Deficiencies or excesses in the cant will induce **lateral forces in curves**. This will bring travelling discomfort for the passengers.
Rails (either long wavelengths[22] or of short wavelength)	A rail may not be absolutely straight when it comes from the manufacturer and the railhead top surface may not be absolutely plane. Due to the train traffic, short wavelength periodic irregularities (corrugations) may develop on the railhead, which generates high-frequency **vibration** of the rail and the wheel, and **noise** is emitted. This will bring travelling discomfort for the passengers.
Wheel out-of-roundness (either long wavelengths[23] or of short wavelength)	Wheel out-of-roundness is one source generating **train and track vibrations**. Especially, short wavelength irregularities often appear on block-braked wheels. This will bring travelling discomfort for the passengers.

Conclusion:

- **As speed increases, the greater the need for high quality and small faults**
- **As the requirements for high comfort increase, the greater the need for residual faults have a large wavelength**

[22] So-called polygonalisation with one to five wavelengths around the wheel.
[23] As above.

17.6 Railhead Irregularities

Railhead irregularities appear on almost every kind of railway track. The development of irregularities on the railhead is explained by the dynamics of the train – track system.

17.6.1 Classification of Railhead Irregularities

The railhead irregularities are divided into three categories[24]:

Table 14 Classification of Railhead Irregularities

Category	Wavelength	Amplitude
Corrugation	30 to 80 mm (3 to 8 cm)	a few hundredths of a mm
Short waves	150 to 300 mm (15 to 30 cm)	up to 1 mm
Long waves	up to 2 m	

Only the longer wavelengths can be explained by the manufacturing process.

17.6.2 Consequences of Rail Corrugation

The rail vibrates out of phase with the corrugations[25]. All the forces are being induced into the track because the inertia of the rail is so much less than that of the wheel.

At certain speeds, sleeper resonances will be excited by the corrugation. **Sleepers** may be exposed to fatigue damage due to the cyclic stresses set up in the sleepers in the resonance situation. Also, **rail pads** and **fastenings** may deteriorate due to the many alternating load cycles. Similar are the effects on the ballast which causes **ballast degradation** and **track settlement** with loss of vertical track profile.

What makes the corrugation phenomenon especially remarkable is that there does not seem to be any direct relationship between the train speed and the corrugation wavelength.

The rail corrugation phenomenon will be discussed in par. 19.3.3.

[24] See [40].

[25] Clark, R. A., Dean, P. A., Elkins, J. A., and Newton, S. G., An investigation into the dynamic effects of railway vehicles running on corrugated rails, J. Mech. Eng. Sci., 24, 65 – 76, 1982.

17.6.3 Long Wavelength Irregularities

Long wavelength irregularities are defined here as irregularities of wavelengths of 300 mm or longer (see par. 17.5).

These irregularities may be

- geometric irregularities on the wheel, or
- geometric irregularities in the track, or
- irregularities of the track stiffness.

17.6.3.1 Wheel Out-of-Roundness

[32] An out-of-round wheel (Figure 57) will excite the train – track structure with frequencies corresponding to the wavelengths of the wheel out-of-roundness.

Fig. 57 Wheel out-of-roundness

17.6.3.2 Rail Manufacturing

The rail manufacturing process may induce rail irregularities. These irregularities will induce a low-frequency excitation of the train and the track.

17.6.3.3 Track Stiffness Irregularities

The wheel will be influenced by the varying stiffness of the track due to sleeper spacing and at switches (turnouts).

Sleeper spacing:

The track is stiffer at the sleepers, and softer in-between, so the wheel will be excited with a frequency f that corresponds to the speed v of the train and the sleeper spacing λ ($f = \frac{v}{\lambda}$).

Switches (turnouts)

The track stiffness varies at switches (turnouts) mainly due to ([32])

- the different lengths of the sleepers
- the different spacings of the sleepers
- the loss of symmetry: the left and right rail will have different stiffnesses (the stock rail keeps the stiffness of the track, whereas the switch rail becomes stiffer because of the longer sleepers supporting that rail (see Figure 58 and Figure 59).

The rail bending stiffness (EI) will change dramatically at the frog (diamond or crossing). Such a sudden change of the stiffness will induce transient and high-frequency vibrations in the train and in the track. The mass is also larger at the frog making inertia forces larger.

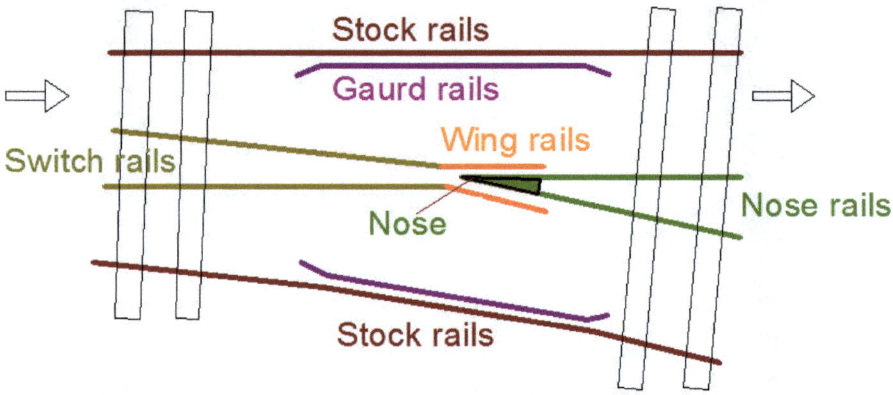

Fig. 58 A railway switch at the frog area

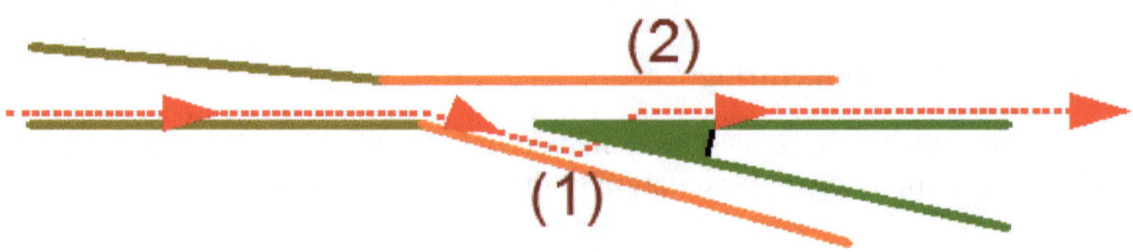

Fig. 59 The wheel path through the frog (diamond or crossing). The wheel leaves the wing rail at point (1) and enters the nose at point (2).

In par. 17.7.3 the impact forces in the switch area are discussed.

Track subgrade
The track superstructure is seldom built on a homogeneous substructure. Due to a variable stiffness of the track subgrade, the track stiffness will vary along the track.

As this stiffness is more or less random along the track, it will induce low-frequency random oscillations of the train.

17.6.3.4 Track Embankment Settlements

When a track is loaded by a train, the ballast and subgrade may undergo (nonelastic) deformations. After the load has passed, the track will not return exactly to its original position but to a position very close to the original one. After thousands and thousands of train passages, all these small non-elastic deformations will give a new track position, varying in different parts of the track [32].

The track position changes over time. The wavelength of these track irregularities will have a magnitude of metres up to hundreds of metres, depending on the subgrade, and will induce low-frequency oscillations of the train. Gradually, the track load variations will increase and so will the track settlement.

Especially, the transition area from an embankment to a bridge is a place where track settlements tend to occur (see par. 15.3.4).

17.7 Impact Loads

Distinct single impact loadings of the track could come from wheel flats or other defects on the wheel tread, rail joints, and switch frog (diamond or crossing) passages. The impact of discrete track defects is further discussed in Chapter 21.

17.7.1 Wheel Flats

Wheel flats (Figure 60) can occur ([32])

- due to problems with the brakes at a wheel (or due to low adhesion),
- when a wheel
    - does not start to rotate immediately as the train starts or
    - the wheel stops rotating before the train comes to a standstill.

In these cases, the wheel slides along the rail for a while. When the wheel slides, part of the wheel tread is removed and a flat surface is formed.

Fig. 60 Wheel flats

Later, when the wheel starts to rotate, an impact load will hit the rail at every rotation of the wheel. This load will bring high frequency vibrations of the track and noise emissions. The impact load could lead -especially at low temperatures- to rail breaks.

The wheel flat causes large dynamic contact forces and stresses also in the sleeper, as described below.

Effect	Description
The flat area of the wheel touches the rail	• The wheel with a flat rotates and the rail touches the flat area of the wheel
Separation of wheel and rail	• At speeds more than 20 km/h (depending on the wheel flat size) the wheel and rail separate for a while, depending on the speed of the vehicle and on flat size, and • the contact force reduces to zero.
The wheel meets again the rail	• When the wheel meets again the rail meet again, a large impact force (P1) is induced between the wheel and the rail (see par. 12.3.2).
Consequences for the vehicle and the track	• The impact forces can lead to fatigue failure of the primary suspension of the vehicle as well as damaging the track components.

17.7.2 Rail Joints

In the case of welded rail joints, there may be a small difference in the vertical position of the two rails connected at the joint. The wheel, "climbs" this difference in level, and especially at high speed large dynamic forces will arise, due to larger mass of the wheel than the rail mass (the wheel will press down the rail, and this will, of course, also create large dynamic forces) [32].

At fishplated joints, or other expansion joints, there could be also a problem of difference in level.

17.7.3 Switches

A switch contains several irregularities both in stiffness and in inertia; the bending stiffness of the switch rail differs from that of the stock rail, the sleepers have different lengths and distances, the frog (the crossing or diamond) is both stiffer (in bending) and has a larger mass than the surrounding rails, and so on.

When a train runs over a switch, dynamic effects appear. As examined in par. 17.6.3.3, there is a change of track stiffness, due to spacing differences and a different length of the sleepers [32].

When a wheel passes the switch blade and the frog (see Figure 59), the wheel will move laterally to find a new equilibrium position, because the wheel – rail contact area will suddenly move laterally on the wheel tread. It induces forces on the wheel and on the track.

At the frog, where the wheel moves over from the wing rail to the nose rail (or vice versa) the geometry of the frog and the wheel profile seldom fit exactly to give a smooth transition. Instead, due to different heights of the wing rail and the

Fig. 61 Cast manganese frog (diamond or crossing)

nose, this transition often induces a vertical impact load on the wheel and on the frog.

Another cause that induces dynamic effects in switches is the sudden change of rail bending stiffness at the frog.

This change applies especially for cast manganese frogs: one large and very stiff block where both the wing rails and the nose rail come together. This block is also extremely heavy and the difference of the masses (block and rails) makes the wheel – rail impact force at the frog (diamond or crossing) even worse.

17.8 Reduction of the Wheel Wear

Wheel wear is the resulting effect of three interacting aspects ([34]):

- vehicle dynamics,
- contact mechanics and
- wheel-rail contact tribology.

To reduce wheel wear we must modify at least one of these aspects.

17.8.1 Wheel-Rail Contact Tribology

There are some ways to improve the wheel-rail contact tribology:

Choice	Comments
Use of materials with improved wear-resistance properties.	Limited improvement due to the fact that a relatively small number of standardised steel grades are allowed for wheel construction.
Use of materials with high wear resistance properties	Probably high sensitivity to RCF damage.
Altering the frictional properties of the wheel-rail interface by the use of friction modifiers.	See par. 17.8.3

17.8.2 Contact Mechanics

Choice	Comments
Use of appropriate wheel profiles for the reduction of local contact stresses and hence of wheel wear.	
Modification of the running behaviour of the entire vehicle	e.g. use of • a special design for the vehicle, or • improved suspension parameters or • advanced technologies like vehicle mechatronics[26].

Any means able to reduce wheel wear will have a significant impact on other fundamental issues related to train-track interaction, especially system performances, and other damaging effects on the wheel and on the rail. This explains why an optimal and universal solution to mitigate wear effects cannot be defined [34].

17.8.3 Use of Friction Modifiers

In curves where unfavourable wheel-rail contact conditions exist, when the flange contacts the gauge corner (see Figure 46), measures are required to reduce the probability of wear occurring to both wheel and rails.

---

[26] Mechatronics is the combination of Mechanical engineering, Electronic engineering, Computer engineering, Software engineering, Control engineering, and Systems Design engineering in order to design, and manufacture useful products.

Rail friction control is being increasingly used to provide this wear mitigation as well as minimise curve noise, lateral forces and corrugations. Application of friction modifiers (FMs), which are basically a composite of inorganic and polymeric materials (in liquid or solid form), provides a thin dry film which, in combination with naturally occurring third body components (primarily iron oxides), gives an intermediate coefficient of friction (about 0,35 [34]). This friction level minimises lateral forces and wear without compromising braking or traction.

17.8.4 Wear-Resistant Wheel Profiles

The contact stresses can be reduced, if for the same contact force, the size of the contact area can be increased, which leads to produce a reduction of the overall wear effect [34].

This consideration has led to the design of the so-called "conformal" or "wear-like" wheel profiles whose shape is designed to achieve conformal contact with the rail, thereby increasing the size of the contact region [34].

Some wheel profiles may present two-point contact with the rail. This increases the wear rate, due to the large longitudinal slippages that are generated by the rolling radius difference between the two points of contact. Therefore, a second way to improve wear resistance is to design the wheel profile so that the formation of two-point contact is avoided.

The design of wear-resistant wheel profiles needs to consider that the shape of the profile will change with the progress of wear. Using wear prediction models it can be ensured that the wheel profile will preserve good wear resistance properties across the entire life before re-profiling (see [34]).

17.8.5 Vehicle Design

A possibility to reduce wheel wear effects is to design the vehicle so as to reduce as much as possible creep forces arising in curves. This in turn can be gained by controlling the angle of attack of the wheel sets (the rotation of the axle with respect to the radial direction in the curve) [34].

Two means can be used to reduce the angle of attack of the wheel sets in curve and hence wear effects associated with curving[27]:

- Decrease of the stiffness of the primary suspension and/or
- shortening the bogie wheelbase[28] (it is characterized by articulated bogies).

The decrease of primary suspension stiffness and of bogie wheelbase both have a negative impact on vehicle stability, since the "wheel-set joining" effect provided by the bogie is lowered. Therefore, the mechanical design of the railway bogie can be viewed as the research of a trade-off between stability and curving

[27] With the exception of active vehicle control (see par. 17.8.6).

[28] The wheelbase is the distance between the front-most and rear-most wheel sets.

requirements. The optimal solution depends on the type of vehicle considered (high-speed, conventional rail, freight transport) and on the service profile specified for the vehicle.

The problem of defining optimal values for bogie parameters can be set in the form of a constrained optimisation problem, in which wear resistance is optimised under the constraint of the vehicle ensuring a required region of stable running [34].

Articulated bogies have a special design in which mechanical links introduce additional degrees of freedom to favour the radial alignment of the wheel sets in curve. Wheel set movements that do not have a role in the curving are constrained to avoid compromising too much vehicle stability.

Articulated bogies are associated with decrease of the critical speed. So, for passenger trains the use of articulated bogies is less attractive. In the case of freight trains the use of articulated bogies provides significant wear reduction.

17.8.6 Active Control of Vehicle Dynamics

The primary suspension design for a railway vehicle imposes a trade-off between curving and stability.

In recent years, there have been proposals to use active suspensions to improve vehicle dynamics, allowing not only an increase in vehicle running performances, but also a reduction in the aggressiveness of the vehicle with respect to the infrastructure and to the vehicle itself. Although these proposals for a "mechatronic train" [29] are still confined to the research field and have not yet found applications, the subject surely deserves a brief examination to underline the possible impact of new technologies in the mid-term future.

The different implementations to actively control the dynamics of a railway vehicle can be grouped into the following categories [34]:

- **active primary suspension**. In this concept, the passive primary suspension is replaced in part or entirely by an actuation system able to apply control forces on the axles. Active control can be used to improve stability, curving and/or to provide guidance.
- **independently driven wheels**. The control action is represented by independent traction/braking torques applied to the two wheels of the axle, providing guidance and improving curving performances.
- **active secondary suspension**. This concept is based on applying a yaw torque on the bogie, rather than on the axles, which can be used to improve stability and/or curving performances.
- **active wheel steering**. In this concept the wheels of the railway vehicle are independently rotated and steered by actuators to provide stability and guidance.

[29] Mechatronics is the combination of Mechanical engineering, Electronic engineering, Computer engineering, Software engineering, Control engineering, and Systems Design engineering in order to design, and manufacture useful products, see par. 17.8.2.

17.9 Maintenance of the Wheel-Rail Interface

(base reference: [34])

Fig. 62 Hatfield derailment[30]

17.9.1 Introduction

Maintenance of the wheel-rail interface aims to

(1) improve safety (avoidance, detection and/or treatment of cracks (at the rail surface or internally) that could lead to a rail break).
(2) reduce costs (improvement of wheel and rail profiles, both longitudinally and transversely).

The two fundamental components of maintenance of the wheel-rail interface are control of friction and control of the wheel and rail profiles.

A well-maintained railway is also one that operates more economically. The consequence of one example of inadequate rail maintenance on a mixed passenger railway is the case of the derailment of a passenger train at Hatfield in England in October 2000 (Figure 62). This is one of the best-known examples of catastrophic failure as a result of inadequate maintenance.

17.9.2 The Importance of Friction

Friction is the resistance (force) offered to the movement of one body past another body with which it is in contact (wheel on rail). Without friction the wheels of a

[30] On 17 October 2000 four people were killed and 70 injured as a result of a derailment near Hatfield.

locomotive could not "grip" the rails. For a locomotive a minimum of friction is desired; Lubrication is important in minimizing friction.

The running wheel induces on the wheel-rail contact a tangential force. The coefficient of friction limits this tangential force. If the coefficient is too low, this limits the traction and (more importantly) braking rates of a vehicle. A low coefficient also limits the tangential force in curving.

It is important to have a sufficient high coefficient of friction. In many countries during autumn, when leaves fall on the rails, the coefficient of friction is reduced to levels of 0,1, hence provoking disturbances in the train operation. On the other side an excess of friction can cause most types of surface damage of rails and wheels, due to the fact that an excess of friction produces heat.

Traction ratios of greater than about 0,4 promote surface damage of wheels and rails, and the coefficient of friction should accordingly be less than this level [34].

Acceleration and braking require an appropriate level of friction (usually a coefficient of friction of no more than 0,2). A certain level of friction is also required to sustain the tangential forces that are required for vehicle stability. Reducing wheel-rail friction can lead to a bogie sliding around the curve.

On sharp curves, some sliding between wheel and rail is almost unavoidable. In order to reduce wear and, accordingly, maintenance, lubrication should be applied to one or both of the sliding surfaces.

17.9.3 Techniques to Maintain the Rail Profile

17.9.3.1 Rail Grinding

One of the important aspects in wheel/rail interface management is the introduction of rail grinding [52]. Rail grinding has been used since the 1980s in maintaining optimal rail profile as well as the elimination of surface defects such as corrugations and head-checks.

The rail grinding process has progressed from a technique for removing corrugations to a high-tech multi-purpose maintenance technique. Rail profile correction, noise reduction and removal of contact fatigue induced head-checks are some of the main reasons for grinding the rails.

17.9.3.2 Grinding Modes

There are three different types of grinding modes when grinding is used to correct rail surface conditions [52]:

- Preparative: cleaning mill scale and construction damage from newly-laid rail to assure optimal initial conditions for commercial service
- Preventive: removing fatigued metal before micro-cracking leads to more serious damage
- Curative/corrective: recovering rail that has been damaged by ballast imprints or other foreign bodies pressed into wheels etc.

By adopting preventive grinding, a surface free of cracks with fresh, work-hardened material is left behind, the overall volume of removed material can be reduced and rail life prolonged. The preventive grinding mode has the advantage to be well planned as rail breaks will not be as frequent as before. Another advantage with preventive grinding is that it requires machines with fewer grinding stones as less material is to be removed. The cost for the grinder can therefore be reduced. A preventive grinding strategy could increase the rail life by up to three times compared to that of a corrective ground rail [52].

A corrective grinding mode removes the work-hardened layer leaving behind a surface with soft steel and with some residual cracks still present due to the fact that the deepest cracks could not be removed.

17.9.3.3 Wheels

Safety is the most important reason for maintaining the transverse wheel profile. "Thin flange" is a safety- critical criterion.

A second safety-critical criterion that dictates the need for wheel maintenance is flange height. If the tread wears sufficiently, thereby essentially increasing the flange height, the flange can eventually interfere with trackwork such as joint bars. This condition must clearly be avoided, which leads to a limit on flange height of about 36,5 mm for common wheel profiles [52].

Section IV: The Track Degradation

In this section track deterioration mechanisms are discussed, which lead to a degradation of the track, and also the factors influencing the durability of the track geometry. Recommendations and good practices for preventing track degradation are also given.

18 The Mechanism of Track Faults Creation

Key terms:

Deterioration, degradation, dynamic effects, rail damage mechanism, rolling contact fatigue (RC), rail corrugation, deterioration of sleepers and fastenings, ballast degradation, subgrade degradation, ballast pockets, soft track, irregularities-inhomogenities, hollows (voids) under the sleepers, track settlements, track defects as cause of track deterioration, drainage, deterioration model,

18.1 Deterioration Processes in Brief

Deterioration or degradation is the *reduction of the original quality due to various influences.*

By far the most significant factor contributing to the deterioration is the dynamic load. The dynamic load is directly related to the axle load and track geometry.

The main processes of track deterioration are

- wear,
- fatigue and
- settlement.

The mechanism of the track faults creation can be presented briefly and simplified as follows (Figure 63):

- Track irregularities (track geometry or running surface) ❶ lead to increase of the dynamic forces, which accelerates the rail wear. Through the fasteners ❷, the consequences are transferred to the sleepers ❸ and the ballast ❹ and finally to the substructure ❺.

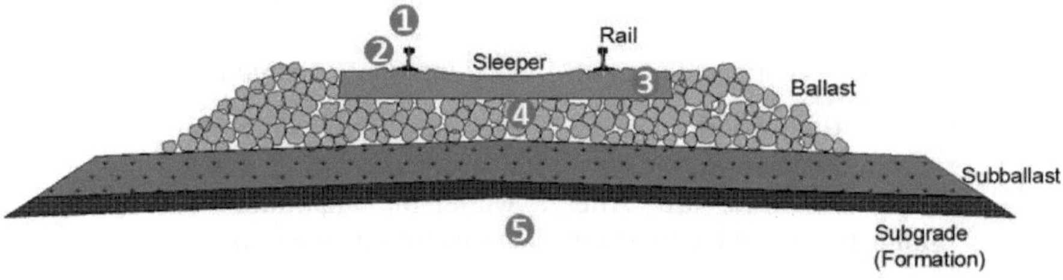

Fig. 63 Degradation process of railways infrastructure

K. Tzanakakis: *The Railway Track and Its Long Term Behaviour*, STTT 2, pp. 115–120.
DOI: 10.1007/978-3-642-36051-0_18

- As a consequence, fine material from the substructure goes up to the ballast.
- The existence or entrance of water to the ballast bed accelerates the deterioration procedure of the track.

18.2 Factors Influencing the Railways Infrastructure Deterioration

Three main groups of factors may be distinguished that contribute to the deterioration of railway infrastructure:

- **Use:** wear by physical contact, static and dynamic load
- **Environment**: climatic influence, water
- **Failures:** faulty components, bad construction

In most of the cases it is not just one of these factors that causes deterioration, but a combination of them.

The bad state or quality of track components contributes significantly to the deterioration of the track geometry. Bad track geometry produces higher dynamic loads on the track components, which accelerates their degradation.

The degradation curve of a generic railways track component is shown by the figure below.

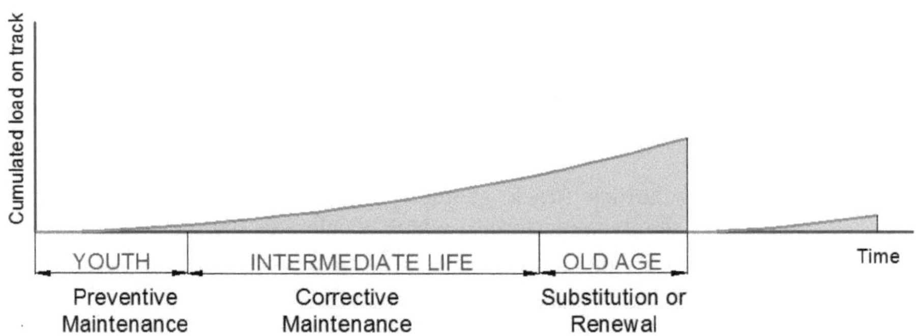

Fig. 64 Degradation processes of railways infrastructure with a load-dependent wearing

The life of such components can be divided into three distinctive parts:

- the **youth period:** During the youth, the component experiences substantial deterioration due to track settlements.
- the **intermediate life:** After a while, the importance of degradations diminishes and the component starts its intermediate life.

- the **old age period:** When reaching the end of its lifetime, the component undergoes higher degradations.

Each period of a component life may be linked to a specific type of maintenance action:

- **Preventive maintenance:** Just after the construction or a complete renewal, preventive works are executed to eliminate any defects resulting from bad construction, faulty material etc. This type of work is intended to avoid any premature degradation of the infrastructure.
- **Corrective maintenance:** Later in the infrastructure lifetime, periodical works are executed to restore the infrastructure to a predefined quality level. Referring to the track as the main component of railway infrastructure, this work type includes corrections of geometry, partly exchange of used or faulty material. This work aims to assure the safety of train operations and to slow down the deterioration process of the track.
- **Substitution or renewal:** At a certain moment these periodical works are no longer applicable, due to either technical limitations or economic reasons. The component has to be partly or completely substituted, if the track no longer fulfils the quality level or requires extremely high maintenance expenses to achieve the quality level.

Track maintenance is discussed in Chapter 32.

18.3 Parameters Affecting Track Deterioration

18.3.1 Driving Forces of Deterioration

In some cases, the track can deteriorate without any traffic (e.g. the soil may settle due to the weight of the embankment, especially in the early years after construction). In most cases running trains is the driving force of deterioration.

A railway track is designed to distribute the loads from trains down to the soil/ground (par. 13.1.3). This distribution through super- and substructure depends on original track design and current track condition. Stiffness of different components in the track structure, as well as the resulting (global) track stiffness will partly determine how the loads are distributed (see Chapter 16).

The static, quasi-static and dynamic forces are all important for track degradation. Some aspects determining forces and thereby influence degradation are listed in the following table [47].

Table 15 Parameters influencing track degradation

Subsystem	Characteristics	Influence of the Subsystem
Vehicle	• speed, • axle load, • unsprung mass, • suspension, • wheel profile, • axle spacings etc.	• static, • quasi-static and • dynamic forces
	• Wheel (current condition such as wheel flats and wheel corrugation)	• dynamic forces
Track	• Track design geometry (curves etc.)	• static and • quasi-static forces
	• Track geometry quality	• dynamic forces
	• Corrugation[31]	• high-frequency forces
	• Rail imperfections such as joints or poor welds	• impact forces

18.3.2 Dynamic Effects

Traffic on rail lines usually comprises many different vehicle types: high speed and conventional passenger trains and also freight trains.

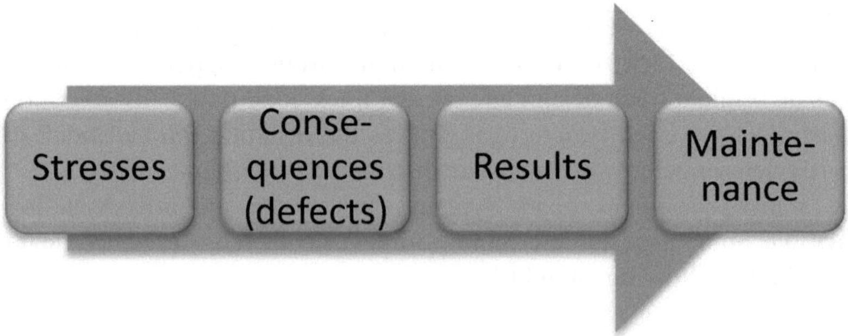

Fig. 65 The track degradation chain

[31] Often called rail roughness.

Track is therefore subject to a wide range of bearing and bending stresses in the rails, pads, fasteners, sleepers, ballast and subgrade.

These **stresses** come about [1]

- the static mass of a vehicle, its wheel sets and the cargo (freight or passenger),
- dynamic actions such as
  - lateral centrifugal forces on curves,
  - longitudinal acceleration and braking forces,
  - rocking of the vehicle about 3 axes (roll, pitch and yaw) (Figure 66),
  - vertical inertial forces from the motion of the wheel set and its suspension,
  - vibrational forces induced from imperfections in the rail surface (corrugations, joints, welds, defects) and in the wheels (flats and shells)
  - dynamic response of the track components to these actions.

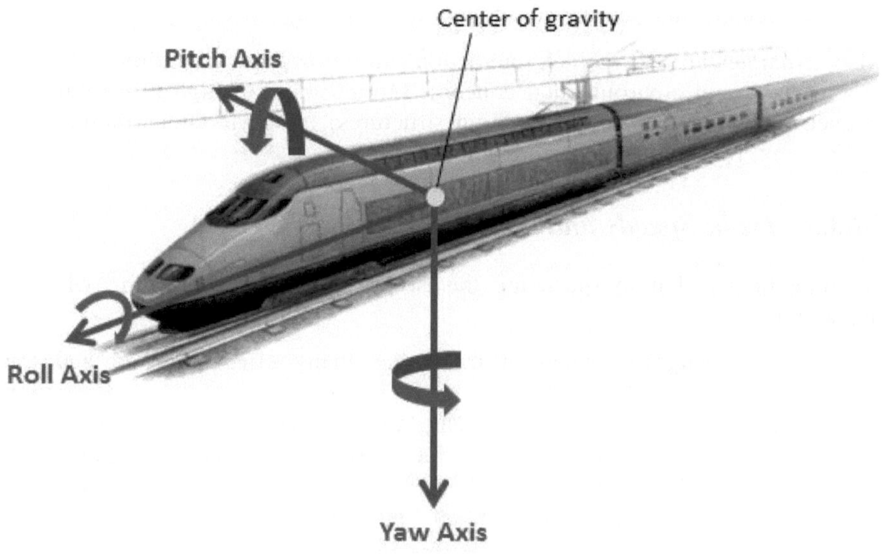

Fig. 66 Rocking of the vehicle about 3 axes (roll, pitch and yaw)

The **consequences** of the frequently large forces generated by these actions are many and wide-ranging:

- Fatigue cracking in rails,
- plastic flow or shelling out of the rail head,
- uneven wear of the rail head,
- cracking or splitting of sleepers,
- loosening of fasteners,

- ballast damage, and
- variations in track alignments and gauge.

Such effects **result** in

- poor riding quality,
- reduced train speed,
- increased fuel consumption,
- potential derailment,
- increased maintenance,
- delays and reduced level of service
- loss of revenue in the longer term.

The key to reducing track **maintenance** demands are the dynamic forces and stresses. Main maintenance actions are:

- rail grinding (to reduce corrugations, and reform the worn rail head to a the correct shape) and
- ballast tamping (to re-establish track geometry in order to reduce the vehicle motions and consequent forces on the track).

Also, dynamics of vehicle/track interaction cause **noise** and **vibration** which have become issues of environmental concern. Minimisation of high (noise) and low frequency (disturbance of surrounding structures) force vibrations therefore has engineering, financial and societal benefits.

18.3.3 Train Speeds and Axle Loads

[1] Axle load and train speed are the two main factors controlling wheel/rail impact forces:

- Increasing the speed of trains has many effects on rail and track behaviour and degradation.
- Axle loads affect rail head wear, fatigue of the rail steel, and plastic flow and shelling in the running surface between wheel and rail head. Greater axle loads lead to increased wheel wear and higher bending stresses in rails and sleepers, as well as higher bearing stresses in ballast and subgrade.

19 The Deterioration of the Track Components

19.1 Introduction

The track is composed of several components as presented in previous paragraphs. Each component has a specific function. The track experiences vertical, horizontal, and longitudinal forces, as described in Chapter 12.

These forces influence the functions of the basic components in the track which in turn affect degradation and the failure process. The failure of each component has an effect on the function of other components in the track system (Figure 67).

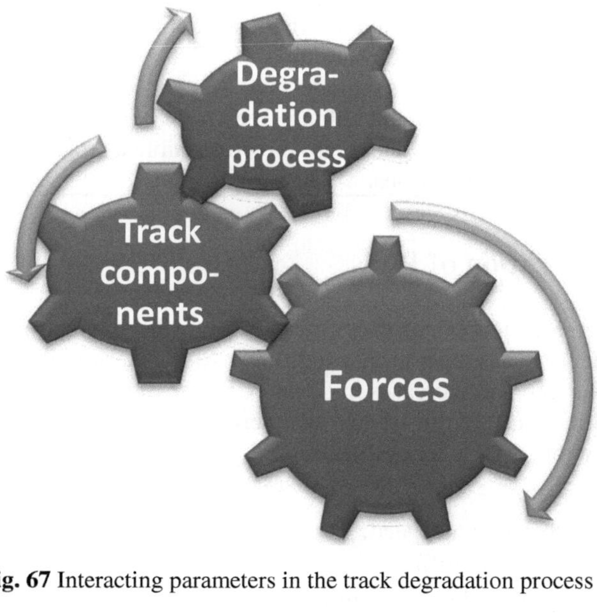

Fig. 67 Interacting parameters in the track degradation process

In order to be able to measure the defects and failures and to prevent catastrophic failures, it is important to analyse the characteristics of the each component which are significant for the deterioration process.

The deterioration process of the different superstructure components is described below:

Track component	Deterioration process
The rails	The rail experience wear and fatigue. Rail deterioration is often a complicated relation between

K. Tzanakakis: *The Railway Track and Its Long Term Behaviour*, STTT 2, pp. 121–145.
DOI: 10.1007/978-3-642-36051-0_19

Track component	Deterioration process
	• vehicle characteristics, • creepage in the wheel-rail contact and • other factors. Deterioration by Rolling Contact Fatigue (RCF) is in many cases significant and also safety-related (RCF initiates cracks that will grow by time and traffic, what can lead to a rail break).
The fastener, rail pads, sleepers	The fastener, rail pads and sleepers have all their own deterioration processes, which are dependent of the forces, bending moments etc. acting on them.
The ballast and sub-ballast	Due to stone reorientation the ballast and sub-ballast layers get differential settlement. This leads to wear of the ballast stones. Fines produced by this wear together with possible other material (ballast fouling), are causing decreased drainage conditions. The subgrade may be exposed to settlement. Varying settlement along the track (resulting in differential settlements), are directly visible as track geometry quality variations.

19.2 Deterioration of Rails

Rail is mainly affected by surface defects due to the interaction between wheel and rail under various operation conditions. Rail deterioration will result in a rougher rail surface and increases in dynamic forces of high frequencies [47].

Wear is the result of the wheel/rail interaction and reduces rail life. Wear is affected by factors, as

- axle load,
- traffic density and speed,
- wheel and rail profile,
- material properties and hardness,
- curvature and superelevation
- grinding practice,
- lubrication and
- rail condition.

A further analysis is given below.

Main categories of rail defects are:

(1) **Abrasive wear** is loss of metal from the side of the rail head by abrasion which occurs mostly in curves (due to contact between the side of the flange of a wheel and the gauge face of the rail) (Figure 45 in Chapter 17).

(2) **Plastic flow or deformation** in rails occurs mostly due to heavy axle load pressure on the rail (the metal of the rail in the wheel and rail contact area become plastic and the material is forced out sideways).

On straight track, the longitudinal forces between the wheel and the rail cause the surface metal to flow along the rail.

In sharp curves, the combination of the contact stress and tangential forces between the wheel and the rail cause the metal to flow sideways on the surface of the rail head.

(3) **Rail corrugation** is actually a type of plastic flow. It influences comfort, noise, loading and fatigue. When the wheel passes over a corrugated rail head[32], the interaction between wheel and rail produces a very high noise level, as well as very high frequency vibrations of the rail. These lead to large dynamic wheel-rail contact forces and can damage the fastenings, Rail pads and sleepers.

(4) **Rail fatigue cracking** is caused mainly by a variety of stress concentration due to corrosion and mechanical damage, because of defects in rails from the manufacturing process and welds due to the passage of traffic.

(5) **Rail creep** occurs when the rail move longitudinally through the fastenings due to traction and braking forces and cause geometry deviations: the rails move up in some sections causing in summer a risk of rail buckling.

The UIC Leaflet 712 [82] harmonizes the methods used for identifying, notifying and classifying rail defects. It includes:

- a definition of rail defects,
- a definition of the recommendations,
- a definition of the location of the defects,
- the general coding system for rail defects.

For each type of broken, cracked or damaged rail there is a page giving one or more photographs of the defect in question and an explanatory text detailing, for each defect:

- its characteristics and appearance,
- means of detection,
- recommendations of a general nature that each railway can adapt to its own specific requirements.

An in-depth analysis of the factors influencing rail degradation is given in [90]. The identified factors responsible for rail degradation are illustrated using a cause and effect diagram (Figure 68).

[32] Particularly short wave corrugations.

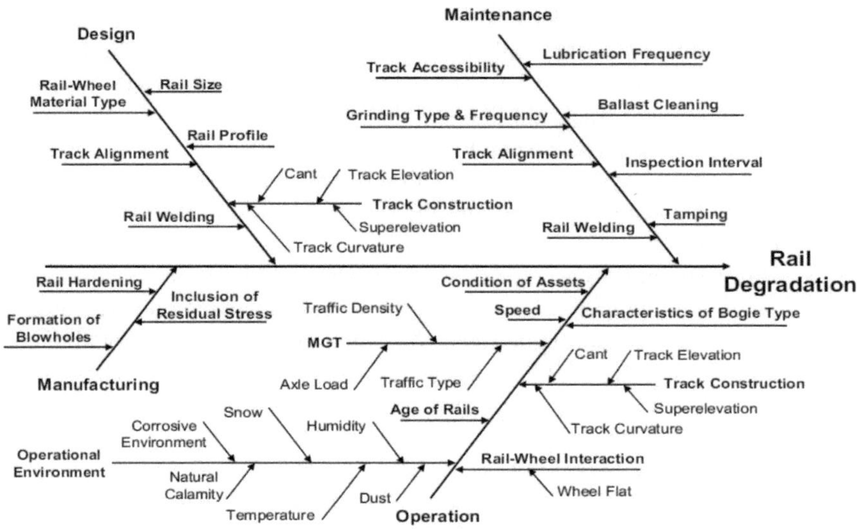

Fig. 68 Cause and effect diagram for the factors influencing rail degradation [90]

19.3 The Rail Damage Mechanism

(base reference: [89])

19.3.1 Crack Development Process

The crack development process consists of three phases:

- crack initiation,
- crack propagation and
- rail break

Crack initiation: When the wheel passes over a rail section, the axial load continues to increase and decrease cyclically. This process continues for a certain number of cycles until plastic deformation begins, usually at a relatively weak location. These deformations take place as a result of compressive and shear forces.

This process goes on for a few cycles until a micro-crack is formed in the rail surface. These micro-cracks develop into cracks when stress is repeatedly applied (micro-cracks may not always develop into cracks).

Once a crack is initiated, stress concentration occurs at both ends of the crack causing it to continue to grow. Cracks initiate in a very thin surface layer of the rail and develop inside the rail head.

Crack propagation: The crack propagation depends on the rail material, the point of crack initiation and the kind of metallurgical processing or the heat treatment method adopted for that particular rail section. If the crack propagation is in an upward direction towards the rail surface, pieces of rail material detach from the rail surface, but if these cracks propagate downwards, they may cause a rail break.

Crack propagation rate is accelerated by fluid entrapment. The presence of water, snow or lubricant on the rails may increase the crack propagation rate. When these small head checks are filled with water or lubricants they do not dry up easily. During wheel-rail contact, these liquids get trapped in the crack cavities and build up very high localized pressure, which may even be greater than the compressive stress. If head checks (see par. 19.3.2.3) are in the direction of train traffic, crack growth takes place due to liquid entrapment, but when head checks are in the opposite direction of train traffic, the liquid is forced out before its entrapment.

Rail break

Rail break is the final result of the crack development process. The first two phases of crack development are critical for railway engineers, as it is during this phase that the crack should be detected by the inspection techniques and consequently rectified by the suitable implementation of maintenance or replacement measures.

Fig. 69 Broken Rails

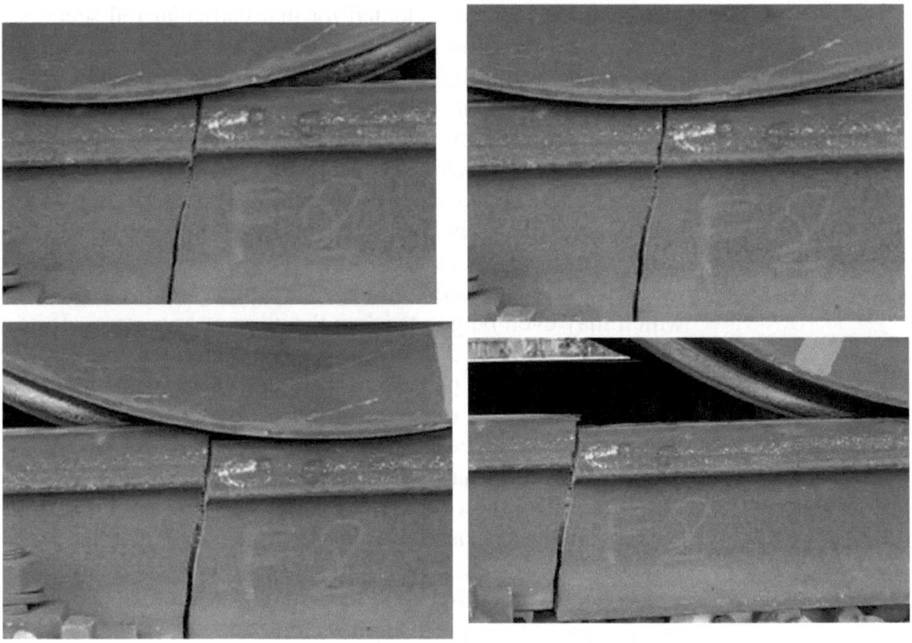

Fig. 70 A vehicle passing a broken rail

19.3.2 Main Rail Problems from Rolling Contact Fatigue (RCF)

19.3.2.1 Causes of RCF

Figure 71 below illustrates the inter-relationship of rolling contact fatigue for both wheel and rail.

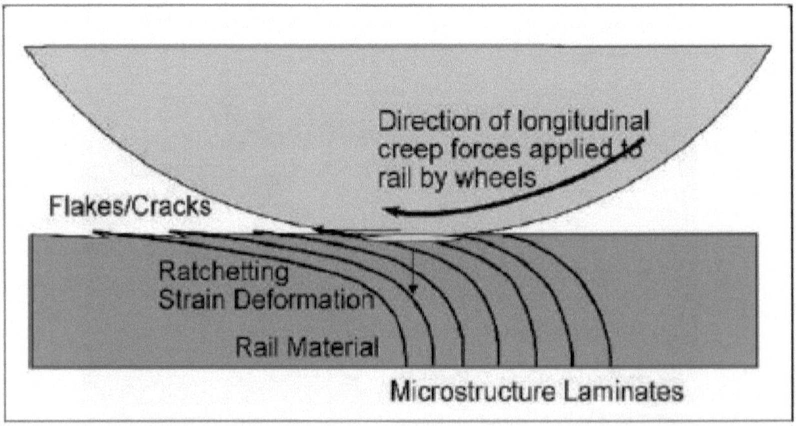

Fig. 71 Explanation of rolling contact fatigue (RCF) ([24][24.3])

The figure illustrates the large shear plastic deformation developed at, and close to, the surface of the material in the contact patch. Cracks will initiate due exhaustion of the ductile of the material in this region under the repeated combined application of surface shear forces and high compressive loads[33].

The orientation of the cracks on the surface will be at right angles to the direction of the resultant (of longitudinal and lateral) creep force. Cracks therefore initiate as a result of the accumulation of strain energy in the material from the forces generated at the wheel/rail interface. These forces are generated in both the vertical and shear (longitudinal and lateral) directions and arise due to the steering behaviour of the vehicle's wheel sets.

The most typical defects due to RCF are squats and head checks.

For minimum actions see [24][24.7].

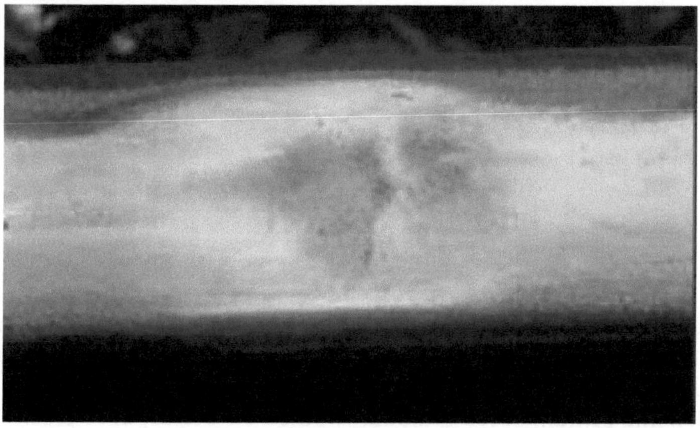

Fig. 72 Squats

19.3.2.2 Squats

Squats are a form of rolling contact fatigue (RCF) that appears as a discrete defect. Squats can be observed visually during track walking or ultrasonic equipped trains can find their location. In most cases after discovery, they will then be inspected manually using ultrasound by trained personnel.

The characteristics of squats, according to UIC Leaflet 712R [7] (defect type 227) are: Squats is visible on the rail head as widening and a localised depression of contact surface, accompanied by dark spot containing cracks with circular arc of V shape. The cracks propagate inside the rail head, at first at a shallow angle to the surface. Then when they reach 3 – 5 mm depth, they propagate downward transversely.

Squat occurrence can be related to track short wave irregularities, especially on the rail top such as indentations, weld and corrugations. Material strength, unsprung mass, traction and braking, sleeper spacing and fastening system properties also play important roles.

[33] This process is known as "ratchetting".

The postulated squats growth process is described in [24][24.5].

19.3.2.3 Head Checks

Head checks are a form of Rolling Contact Fatigue that appears on rail as a series of cracks along the running band inclined to the gauge corner, Figure 73.

Fig. 73 Head Checks

Head checks are initially found by visual inspection or by eddy current systems mounted on trains. Once found a follow up investigation is carried out by manual ultrasonic and/or eddy current inspection.

The planning of train based grinding is based on the depth of cracks found. In all cases after the cracks have extended a certain distance grinding is no longer permissible and rail replacement is the only option.

19.3.3 Rail Corrugations

19.3.3.1 Causes for Rail Corrugations

Rail corrugation is a surface defect of the rail, which is manifested as periodic wear. If the defect is not removed, the corrugation may lead to high levels of wheel / rail noise and crack initiation. There are many different types of corrugation, some of which are well understood; but the causes and consequent solutions are poorly understood for corrugations in sharp curves, where corrugation is exhibited on the low rail [24][24.3].

19.3.3.2 Potential Mechanisms

The mechanism of corrugation can always be defined as the result of both dynamic and structural factors: for corrugation to occur a wavelength fixing mechanism and damage mechanism is required [24][24.3].

The cause of the periodicity will be a function of track geometry and mechanical characteristics, axle mode, curving behaviour of bogie and vehicle speed. It is also expected that tangential forces and creep will contribute to the damage.

Studies ([91][92]) have looked at Rail pad softness and found that softer Rail pads remove at least one of the corrugation formation wavelengths. Soft pads decouple the rail from the dynamic behaviours below the rail pad, e.g. the transmission of sleeper resonances back to the rail head.

19.3.3.3 Current Solutions

Current practice consists predominantly of using lubrication and friction management and also grinding regimes as corrective action by removing existing corrugations, and preventive action by removing other defects that may trigger dynamic loads and restoring optimum wheel rail contact condition. Harder steel rails may be used on curves to also reduce wear [24][24.3].

19.3.3.3.1 Lubrication and Friction Management

Lubrication is used on curved tracks to reduce the friction and wear that occurs between the wheel flange and the gauge side of the rail (Figure 46, see also par. 17.4). Lubrication slows the initiation of surface and contact fatigue cracks but has adverse effect on crack propagation.

The coefficient of friction between wheel and rail varies in track between a minimum of almost 0 and a maximum of up to 0,8 ([24][24.8]). It is widely accepted that the ideal coefficient of friction μ in the contact area between wheel and rail on the running surface should be between 0,30 and 0,35, whereas the respective value at the flange / gauge corner contact should be a minimum. Providing consistently such desirable friction values would result in reduced abrasive wear, reduced development of RCF and short wave corrugation in curves as well as improved steering performance of vehicles, energy savings (e.g. a 5% reduction is reported from heavy haul railways in North America) and reduced noise levels. The positive effects might differ depending on the type of traffic (heavy haul, mixed traffic, light rail…).

Lubricants were developed initially to reduce the coefficient of friction at the flange / gauge corner to a minimum. These substances are either applied locally in track with stationary equipment (greasing the gauge face of high rails in curves) or using vehicle based equipment. Gauge face lubrication of high rails in tighter curves has become a widely used maintenance activity and can be classified as state of-the-art.

Rail mounted lubricators are currently used for the following reasons:

- To reduce the rate of gauge face wear of rails.
- To reduce the traction coefficient and limit the initiation of RCF, in particular gauge corner cracking.
- To help prevent incidence of insulated block joint failure by reducing the risk of metal particle inclusion (insulation break down).
- To reduce the risk of potential derailment due to the combination of dry gauge face/wheel flanges causing high friction and flange climb derailment.

Lubricators shall be fitted at all possible locations where high lateral forces between wheel and rail are known or observed to exist (using the rail wear pattern as a guide)[34]. Such sites would meet the criteria below:

- All curves exhibiting or having had a history of excessive gauge face wear;
- All curves with a history of RCF;
- All curves with a radius of less than 1500 metres;
- Cant deficiency greater than 50 mm;
- All curves with continuous check rails installed due to curvature should also be lubricated for reducing noise and wear.

Lubrication is generally influenced by many factors including wheel and rail profile, surface conditions of the wheel and rail, track geometry, the viscosity and lubricity of the grease, operating temperatures of the wheel and rail, and finally environmental factors such as temperature and precipitation.

Common lubrication methods are:

- Wayside lubrication (Figure 74)
- Locomotive flange lubrication (Figure 75)
- Hi-rail (high rail) based lubrication

Fig. 74 Wayside lubrication

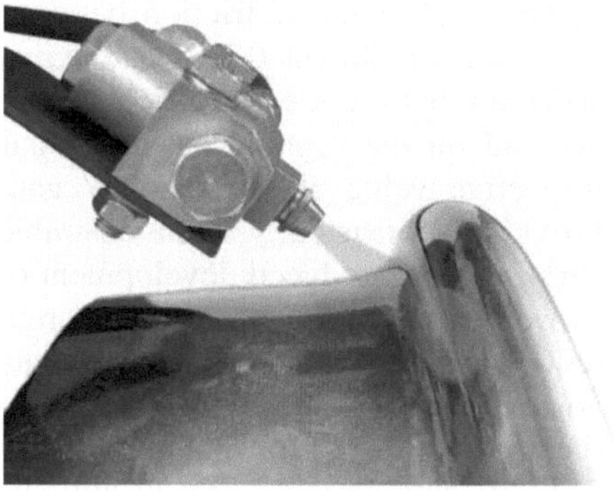

Fig. 75 Locomotive flange lubrication

Proper lubrication can reduce rail/wheel wear rates about 10-15 times in 300-400 meters curve radius and 2-5 times in 600 meter curve radius leading to significant cost savings for both the Infrastructure Manager and the Rolling Stock owner.

19.3.3.4 Rail Grinding

RCF can be effectively controlled when the shape of the rail head profile matches the prevailing wheel profiles and when the fatigued rail surface is regularly removed in appropriate intervals.

---

[34] Network Rail Company Standards.

As a rule grinding should always be carried out as soon as possible after re-railing in order to introduce or optimise the target profile (wheel-adapted or anti head-check (AHC) profiles). In order to assure a long rail service life and consequently low LCC the consecutive grinding cycles need to be programmed from the beginning [24][24.9].

19.4 Deterioration of Sleepers and Fastenings

The sleeper and fastening deterioration depends on traffic and operational factors such as axle loads, speeds, accumulative tonnage and maintenance practices.

Concrete sleepers deterioration composes

- sleeper cracking,
- loosening of fastenings when support of the sleeper is inadequate and
- abrasion of the soffit[35] due to excessive movement of sleepers or bad ballast conditions.

A vital factor for the fastenings in track to function successfully is the condition of the sleepers. Deterioration of fastenings such as loose, broken or missing fasteners influences other track components. Also track geometry irregularities arise and other deterioration modes become apparent on track components.

19.5 Ballast Degradation

19.5.1 Overview

The ballast bed consists mainly of the skeletal grain (Figure 76) and -depending on the operating load and track conditions- distance grain and the filling grain (Figure 77, Figure 78).

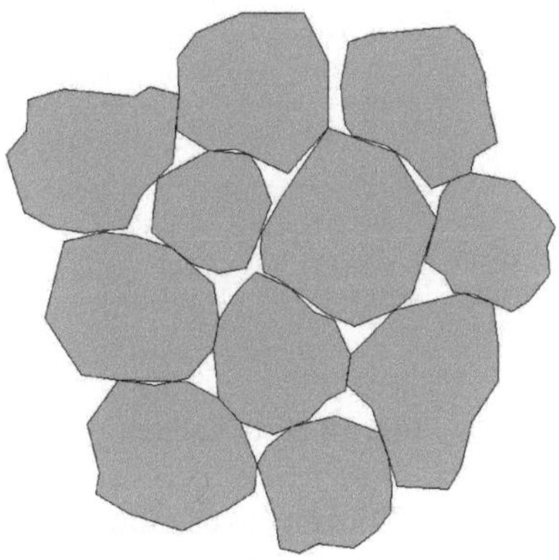

Fig. 76 Clean new ballast (skeletal grains)

[35] The underside or base of sleeper.

A newly produced ballast bed consists almost entirely of the skeletal grain. Due to the operating loads from the trains, the edges of the new ballast will be highly stressed.

The rolling of the wheel on the rail causes just before reaching the sleeper a slight lifting and then by rolling over the sleeper a hammering of the sleeper to the ballast. This leads to crumbling (splintering) of the ballast grains at the contact points (Figure 77). This continues until the contact areas are so numerous, that the loads can be transferred through the gravel to the sub-ballast and the subgrade.

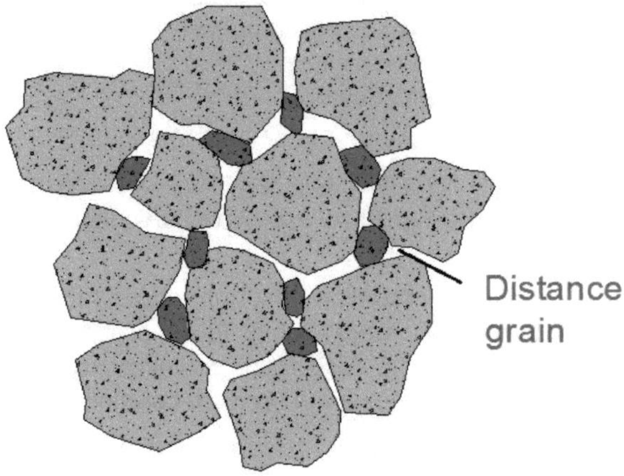

Fig. 77 Ballast after operating loads (skeletal and distance grains)

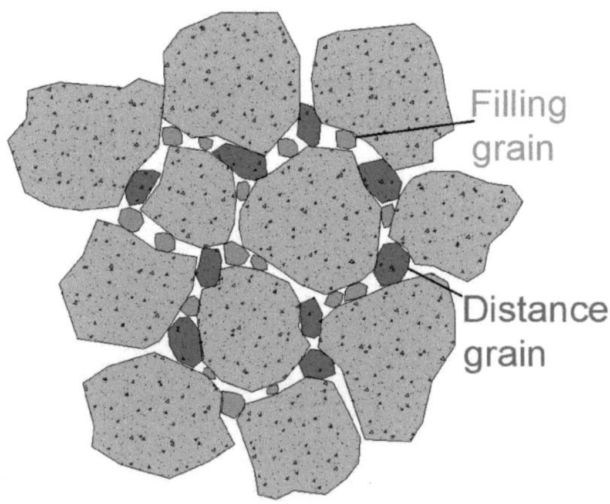

Fig. 78 Dirty ballast

The distance grain (Figure 77), improves the shear strength. Over time, however, the proportion of grain distance, forming an additional finer material, the filling grain that after an operating load of several million cycles encloses the skeletal grains (Figure 78). Thus, the internal friction angle of the material is going to be smaller, the shear strength is reduced and so the carrying capacity. Filling grain can also enter into the ballast bed by rising grains from fine-loamy or clayey subsoil.

Ballast is the main reason of settlement of railway track. Constant settlement along the track does not affect track deterioration. Irregular or differential ballast settlement is one of the reasons for track degradation (Chapter 20).

Fig. 79 Dirty ballast at ballast depot

Fig. 80 Unsuitable ballast in track

Fig. 81 Vegetation in a track with unsuitable ballast

19.6 Subgrade Degradation

The subgrade is another very important component of the track and its stiffness has a significant influence on other track components' performance.

Subgrade defects are closely related to track surface irregularities. Possible defects which are not the direct result of traffic such as clay holes, localised weakness, cut slope erosion due to blocked drainage (Chapter 23), and short shoulder due to end of culvert failure, are difficult to predict.

Traffic related failure modes are generally predictable. They include:

- massive shear failure or through track slide;
- progressive shear failure: shoulder slide and track squeeze; and
- attrition.

19.6.1 Subgrade Failure Modes

Subgrade degradation is mainly due to three factors:

- Repeated traffic loading (affects subgrade stiffness)
- weight of trains and
- environmental factors including fine grained soils and high moisture.

Subgrade degradation occurs in the presence of water. Water at the interface of ballast and subgrade leads to the formation of slum. Due to the cyclic loading by trains, the slum migrates into the ballast which becomes highly fouled. The slurry can block the drainage system, which enhances the subgrade bed weakening.

Figure 82 shows a few factors that contribute to subgrade deterioration:

- repeated dynamic loading,
- excessive moisture and
- fine grain or poor quality soil.

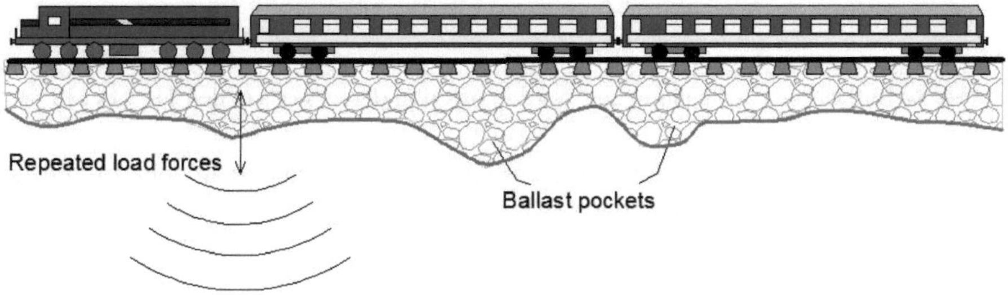

Fig. 82 Subgrade deterioration factors

19.6.2 Ballast Pockets Associated with Common Subgrade Failure Modes

A *ballast pocket* develops when a depression develops in the top formation or subgrade below the tracks.

Figure 83 and Figure 84 illustrate ballast pocket formation as a result of the vertical component of progressive shear deformation. Often this type of failure contributes to the development of non-uniform track settlement and undesirable track geometry changes.

Water trapped in a ballast pocket is a major contributor to many "soft track"[36] situations.

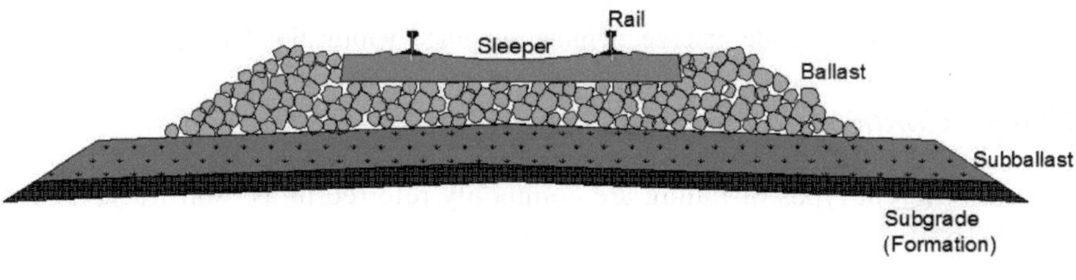

Fig. 83 A typical cross section of a track (without ballast pocket formation)

[36] See par. 19.6.3

In Figure 84 and Figure 85 there are typical sections showing progressive failure of the subgrade as a result of ballast pockets.

In Figure 84 the water is trapped in the ballast pockets, the ground begins to move sideward (in the figures the sub-ballast isn't shown, although similar phenomenon will occur).

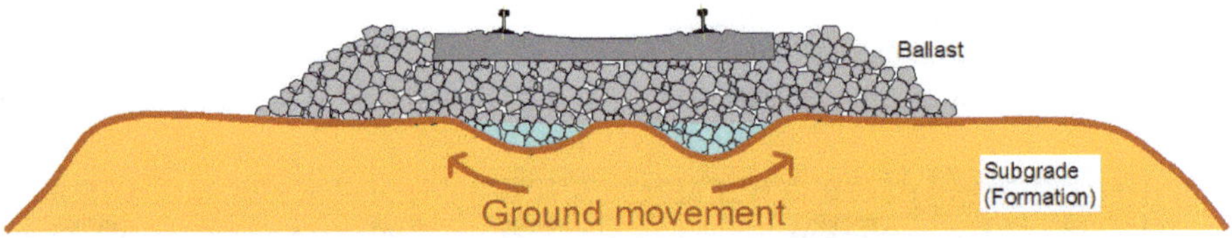

Fig. 84 Ballast pocket development: water trapped in depressions under rail (after [87])

When this phenomenon develops, a subgrade failure can occur (bearing failure with shear displacement). The track is water saturated.

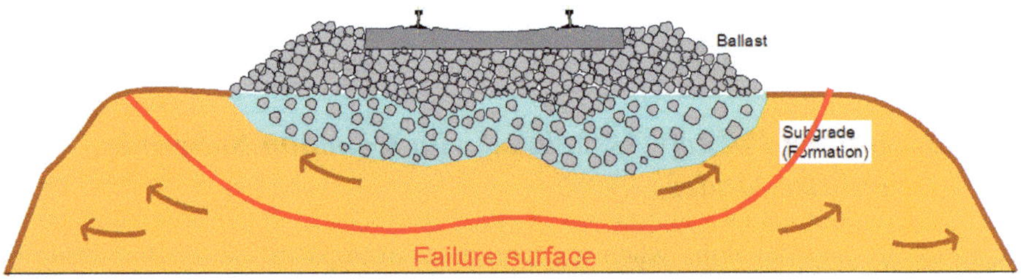

Fig. 85 Bearing failure (with shear displacement). Water saturated track structure (after [87])

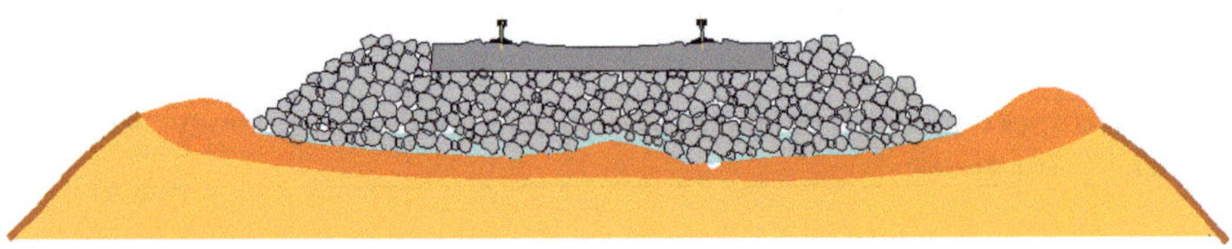

Fig. 86 Subgrade squeeze, minimal subgrade deformation (after [87])

19.6.3 Contributors to "Soft Track"

Several different types of failure are commonly referred to as "soft track". These include [86]:

- Ballast failure and pumping track
- Sub-ballast failure
- Shallow subgrade failures (sometimes caused "track squeezes")
- Rigid layer pumping problems (sometimes called "mud pumping")

- Embankment failure
- Larger ground failure (landslide)

While all of the above can produce cross-level and vertical and horizontal alignment defects, they often have different causes and different treatments. Considerable experience is required to differentiate between these different failure types and to select an appropriate remedial measure to correct the problem. No single remedy has been identified to correct all of these problems; but since most of these failure modes involve water, drainage improvement is almost always part of the remedy. In some situations ballast *undercutting* or shoulder cleaning may be best. However, it is important to recognize that ballast *undercutting* and shoulder cleaning also improve track drainage.

Fig. 87 Soft track [86]

Factors that commonly contribute to subgrade related soft track conditions include ([86] and Chapter 23):

Table 16 Contributors to "Soft Track"

Water	• Rainfall • Water in ballast pockets • Water in ballast • Water between "rigid slabs" and underlying subgrade soils • Water in embankment soils and cracks in embankments • High groundwater and springs • Embankment erosion	

Table 16 (*continued*)

Weak subgrade soils	• Clay subgrade • Silt subgrade • Saturated or wet soils • Progressive shear failure (and build up of ballast pockets)	
Overstressed subgrade soil	• Insufficient ballast and top formation thickness • Progressive shear failure (and build up of ballast pockets) • Increased train loadings • Too high or too steep embankment	
Poor Construction Practices	• Inadequate foundation preparation	• Constructing new fill over topsoil • Constructing new fill over wet, weak, or soft ground • Not benching into existing ground prior to placing fill
	• Inadequate compaction or moisture conditioning of fill • Fills constructed too steep • Inadequate drainage systems • Use of "wrong" type of fill	• Use of oversized rock • Use of moisture-sensitive soils
Poor Maintenance Practices	• Inadequate cess drain and culvert maintenance • Over-steeping of embankments (by excessive track raising or widening of tops of fills) • Placing fill on slopes • Undercutting of slopes	
Fouled ballast		

The other types of subgrade failures are subgrade attrition with mud pumping caused by high moisture content and repeated loading at the ballast and subgrade interface leading to ballast fouling and associated drainage problems, slope stability failures and excessive consolidation settlement due to self-weight.

The most commonly used remedial measure, is to increase the ballast thickness to reduce subgrade stresses. However, this method of remediation in association with excessive plastic deformation can also lead to formation of ballast pockets. This remedial measure appear to provide only a short term solution by only treating the symptoms of the problems rather than analysing its root cause for longer term eradication.

Table 17 presents the major subgrade problems, causes and features.

Table 17 Major subgrade problems causes and features (after [88])

Type	Causes	Features
1. Progressive shear failure	• Repeated over stressing subgrade • Fine-grained soils • High water content	• Squeezing of subgrade into ballast shoulder • Heaves in crib and/or shoulder • Depression under sleepers trapping water
2. Excessive plastic deformation (ballast pocket)	• Repeated loading of subgrade • Soft or loose soils	• Differential subgrade settlement • Ballast pockets
3. Subgrade attrition with mud pumping	• Repeated loading of subgrade stiff hard soil • Contact between ballast and subgrade • Clay-rich rocks or soils • Water presence	• Muddy ballast • Inadequate sub-ballast
4. Softening subgrade surface under Sub-ballast	• Dispersive clay • Water accumulation at soil surface • Repeated train loading	• Reduces sliding resistance of subgrade soil surface
5. Liquefaction	• Repeated dynamic loading • Saturated silt and fine sand • Loose state	• Large track settlement • More severe with vibration • Can happen in sub-ballast
6. Massive shear failure (slope stability)	• Weight of train, track, and subgrade • Inadequate soil strength	• Steep embankment and cut slope • Often triggered by increase in water content

Table 17 (*continued*)

Type	Causes	Features
7. Consolidation settlement	• Embankment weight • Saturated fine-grained soils	• Increased static soil stress as from weight of newly constructed embankment • Fill settles over time
8. Frost action (heave and softening)	• Periodic freezing temperature • Free water • Frost-susceptible soils	• Occurs in winter/spring period • Heave from ice lens formation • Weakens from excess water content on thawing • Rough track surface
9. Swelling/shrinkage	• Highly plastic or expansive soils • Changing moisture content	• Rough track surface • Soil expands as water content increases • Soil changes as water content decreases
10. Slope erosion	• Surface and subsurface water movement • Wind	• Soil washed or blown away • Flow onto track fouls ballast • Flows away from track can undermine track
11. Slope collapse	• Water inundation of very loose soil deposits	• Ground settlement
12. Sliding of side hill fills	• Fills placed across hillsides • Inadequate sliding resistance • Water seeping out of hill or down slope is a major factor	• Transverse movement of track

19.7 Effects of Irregularities – Inhomogenities in the Ballast Track

Inhomogenities in the ballasted track are difficult to avoid or even minimize. Thus, a high output quality can be a highly homogeneous subgrade, a homogeneous ballast layer and rails with a good running surface are reached. It has been shown that the initial (output) quality of a ballasted track is crucial for the development of settlements in the track (see Section V). Each line has a typical profile, which is set during construction due to existing inhomogenities, which will deteriorate again after each tamping [8].

19.7.1 Effects of Pollution in the Ballast Bed

Once the ballast bed is polluted (Figure 88), an unfavourable pressure distribution acts on the subgrade. The pollution caused by rising fines from the surface increases.

In the muddy ballast, friction is greatly reduced. It comes in uneven settlement of the track. The restoration of the track position by "tamping" is only a short time effectively. The track position is deteriorating rapidly; the subgrade is still stressed harder. Clay goes on the surface of the ballast bed. Settlements occur in the track, which can sometimes lead to derailments.

Fig. 88 Polluted ballast bed

19.7.2 The Formation of Hollows (Voids) under the Sleepers

The contact problems between the sleeper and the ballast lead to hollows under the sleepers (voids).

There are different types of contact problems between the sleeper and the ballast [31]:

Central hollow (Figure 89)	A hollow forms at the centre of the sleeper and extends symmetrically in both directions. The void is assumed to expand up until the sleeper becomes fully hung. This case is easily detectable by visual inspection of a poor-quality railway track.

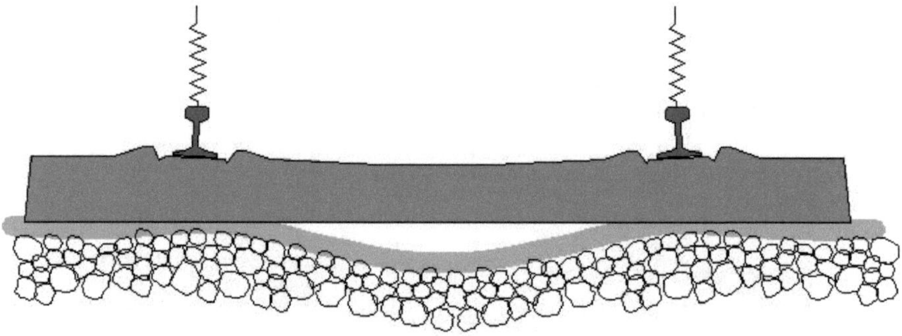

Fig. 89 Sleeper/ballast contact patterns: central hollow

Single hanging (Figure 90)	The hanging condition forms at one end of the sleeper. The hanging space grows incrementally from perfectly full contact to totally hung condition. This void can extend from either end of the sleeper.

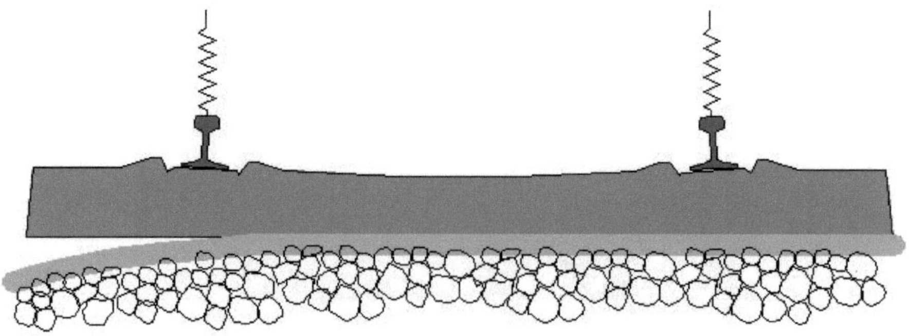

Fig. 90 Sleeper/ballast contact patterns: single hanging

Double-side hanging (Figure 91)	The double-side hanging can also occur when the gap between the sleepers and the ballast forms at both ends. The sleeper/ballast contact in this case remains only in the middle segment of the sleeper.

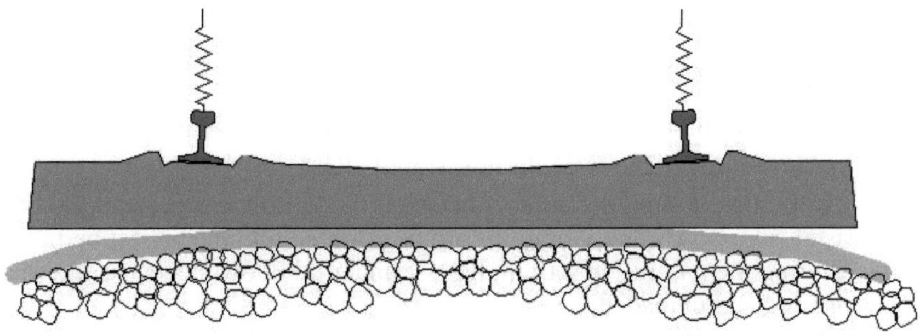

Fig. 91 Sleeper/ballast contact patterns: double hanging

Triple hanging (Figure 92)	In this contact pattern, not only do the hollows format both ends, but also there is a pocket growing up at the centre of the sleeper.

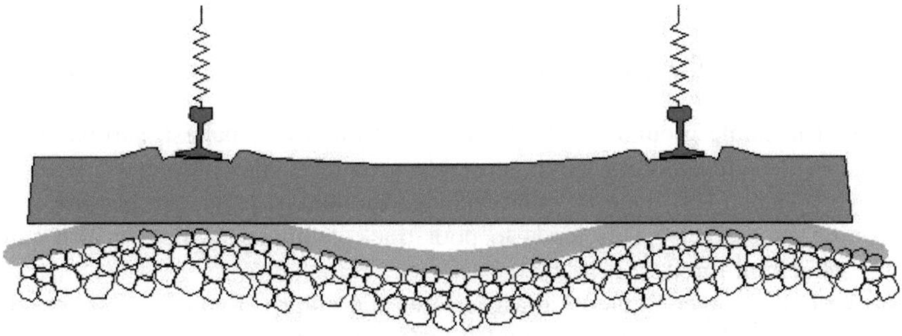

Fig. 92 Sleeper/ballast contact patterns: triple hanging

Side-centre hollows (Figure 93)	The asymmetric contact configurations are considered in this case for which the edge hanging and the central hollow are supposed to form at the same time.

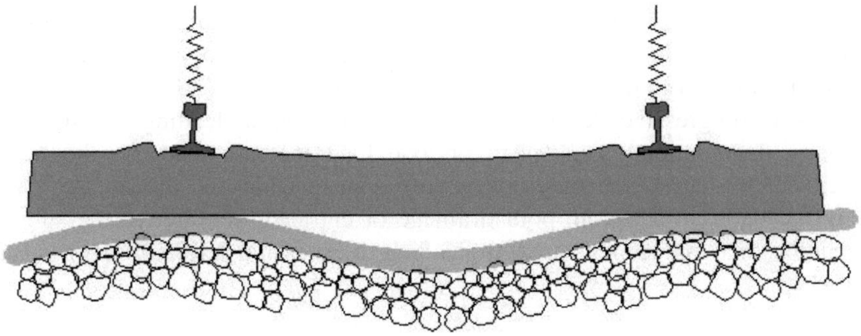

Fig. 93 Sleeper/ballast contact patterns: side-central hollows

A hollow position under the sleeper can develop, as describe bellow.

1. During the track construction process the grain structure of the ballast is stabilized.
2. After the first train passing a certain hollow position distribution is achieved due to the irregularities (rail, ballast, soil).
3. At a train passing, the contact between the hollow-sleeper and the ballast is closed. Loads and vibrations are transmitted into the ballast bed.
4. During the passing of a train, ballast under a hollow-sleeper is alternately loaded and unloaded.
5. Through increased dynamic effects (the sleeper hammers on the ballast) and the intermediate full load-discharge of the ballast (when the sleeper moves upwards), the initial stabilized structure of the ballast grain is reduced in the horizontal direction.
6. Due to abrasion or ballast grain destruction at the position of a hollow sleeper, often tamping is necessary which leads to a stronger ballast pollution and at the end to a track destabilization.

19.7.3 Causes for Unstable Ground (Substructure) – A Summary

Causes of unstable ground can be from any of the layers ballast, sub-ballast and subgrade (formation). Very often, unstable ground is also associated with poor drainage (Chapter 23). Due to poor drainage the ballast layer can become weak, deforming excessively. This leads to poor track geometry, which in turn causes higher wheel/rail forces ([24][24.3]).

In the ballast layer, fouled ballast is often the culprit. Fouled ballast has very high stiffness and little damping, and therefore is not capable of accommodating high wheel/rail forces associated with poor track geometry.

A saturated sub-ballast layer due to fouling materials and poor drainage can lose its strength under repeated dynamic wheel loads, thus becoming an unstable layer.

If formation is built with soft or marginal soils, it can become unstable either in a progressive manner or suddenly. Sudden formation failure rarely occurs, unless there is a dramatic change of environment (such as high rainfall and flooding) or load conditions (such as a large increase of wheel loads). Progressive deformation (shear), however, often occurs for the formation built with soft or marginal soil types. This progressive deformation can become rapid, leading to rapid track geometry degradation when speed or axle load increase, because the bearing capacity of formation soil may not be sufficient to withstand the stresses caused by traffic loads. In addition, poor drainage or entrance of water to the subgrade (formation) may reduce soil strength, leading to excessive deformation or unstable ground.

19.7.4 Techniques for Correcting Unstable Ground

Established techniques for correcting unstable ground include:

[1] Geogrids that can improve bearing capacity of formation
[2] Short piling, geo(stone) piers or lime-cement pillars installed under the ballast layer to improve bearing capacity of formation
[3] Ballast undercutting and shoulder cleaning to improve ballast drainage
[4] Design lift tamping (or stone blowing) to improve ballast deformation characteristics
[5] Sufficient ballast layer thickness with good quality ballast materials to reduce stresses transmitted to the formation
[6] Adequate sub-ballast layer (formation protection layer)
[7] Removal and replacement of poor formation soil
[8] Hot mix asphalt underlay (between ballast and subgrade (formation)) to reduce stresses transmitted to the formation and to prevent surface water penetration into the subgrade (formation).

20 Track Settlements

Railway track will settle due to the permanent deformation in the ballast and underlying soil.

A new track is straight and well levelled. After having been used some time, the track loses its perfect track geometry due to the train traffic. The severity of the settlement depends mainly on the quality and the behaviour of the ballast, the sub-ballast, and the subgrade.

Track settlement takes place in two phases [32]:

- Phase A - Directly after tamping: the track has perfect geometry. The settlement is relatively fast until the gaps between the ballast particles have been reduced and the ballast is consolidated.
- Phase B - After the track has been used for some time: Settlement is slower and there is a more or less linear relationship between settlement and time (or load). This settlement is caused by several basic mechanisms of ballast and subgrade behaviour:

  Compression of ballast and subgrade[35]*:*

  - Continued (after the first phase) volume reduction, i.e., compression of the ballast and subgrade, caused by particle rearrangement.
  - Sub-ballast and/or subgrade penetration into ballast voids. The ballast sinks into the sub-ballast and subgrade. The track geometry changes accordingly.
  - Ballast volume reduction caused by
    - particle breakdown from train loading or environmental factors (ballast particles may split into smaller particles).
    - abrasive wear at points in contact with other particles (i.e., originally cornered stones become rounded, then occupying less space).

  Inelastic behaviour of the ballast and subgrade materials:

  - Inelastic recovery on unloading: Due to micro-slip between ballast particles at loading, all deformations will not be fully recovered upon unloading the track.

[35] The train load may also have an opposite effect: due to the elastic foundation, the train load will lift the track (rails and sleepers) in front of and behind the loading point, thus reducing or eliminating the preload (the dead load) caused by the rails and sleepers on the ballast.

K. Tzanakakis: *The Railway Track and Its Long Term Behaviour*, STTT 2, pp. 147–148.
DOI: 10.1007/978-3-642-36051-0_20

- Movement of ballast and subgrade particles away from under the sleepers causes the sleepers to sink into the ballast and subgrade.
- Lateral, and possibly also longitudinal (in the rail direction) movement of sleepers causing the ballast beneath the sleepers to be “pushed away,” and the sleepers will sink deeper into the ballast.

21 Discrete Track Defects as Cause of Track Deterioration

Wheel/rail contact happens over a very small area (see Chapter 17). In this small area the stresses can cause a variety of defects. Defects or irregularities in either wheel or rail generate sharp peak forces (high frequency forces) in the vehicle and the track (par. 17.7).

Flat wheels and rail corrugations cause periodic excitation. Dipped weld joints and rail shelling and spalling cause non-periodic excitation. The periodic defect is more serious than the non-periodic defect, but the non- periodic defect can produce much higher impact forces.

The range of different defects can be classified as following [10]:

- isolated irregularities in the rail running surface,
- periodic irregularities on the rail surface such as corrugations or repetitive effects of the sleeper spacing,
- wheel defects such as wheel flats and wheel eccentricity, vehicle misalignments,
- random geometrical deviations of the longitudinal profile, and
- random variation in sleeper support stiffness.

The following table gives an overview of the different irregularities, the correspondent types of defects, the forces affected and the related induced stresses (see [10]).

Table 18 Types of defects based on their characteristics (After [10])

Class of Irregularity	Types of Defects	Forces or Loads affected	Stresses Induced
Isolated Irregularities	By default, manufacturing or construction, dipped weld joint, shelling and spalling	P1 + P2 dynamic forces	Contact stress, longitudinal stress and shear stress
Periodic Irregularities	Wear	P2 dynamic forces	Longitudinal stress and shear stress

K. Tzanakakis: *The Railway Track and Its Long Term Behaviour*, STTT 2, pp. 149–150.
DOI: 10.1007/978-3-642-36051-0_21

Table 18 (*continued*)

Wheel Defects	Corrugation, wheel flats and wheel eccentricity	P1 + P2 dynamic forces	Vertical bending stress
Vehicle Misalignments	Hunting and yawing	P2 dynamic forces. and lateral forces	Lateral bending stress
Geometrical Deviations	Track changes in alignment and gauge, and uneven settlement	Quasi-static dynamic forces, P2 dynamic forces and lateral forces	Vertical and lateral bending stresses and bearing stress
Variation in Sleeper Support Stiffness	Voids in ballast under sleeper and fouled ballast	P2 dynamic Forces	Vertical and lateral bending stresses and bearing stress

22 Track Problems, Causes, Mechanisms, Solutions

22.1 Important Track Problems and Their Underlying Causes

Table 19 lists the full range of problems identified and shows how the IMs are affected by each one ([24][24.2]).

Table 19 Problems identified and shows the IMs, listed by frequency of reports

Nr.	Component	Problems	Frequency of Reports
1	Track	bad track geometry	75%
2	Rail	cracks and fatigue	75%
3	S+C[38]	switch wear	75%
4	Substructure	unstable ground	62,5%
5	Joints	insulating joint failure	50%
6	Rail	corrugations	37,5%
7	Rail	wear	37,5%
8	Structures	major line closures	25%
9	Fasteners	worn/missing pads	25%
10	Sleepers	renewal optimisation	25%

[38] Switches and crossings.

K. Tzanakakis: *The Railway Track and Its Long Term Behaviour*, STTT 2, pp. 151–155.
DOI: 10.1007/978-3-642-36051-0_22

Table 19 (*continued*)

11	Culverts/pipes	flooding	25%
12	Ballast	stone spray	25%
13	Ballast	ballast wear	25%
14	Rail	low friction/adhesion	25%
15	Joints	weld quality	25%
16	S+C	common crossings	25%
17	S+C	manganese diamond (crossing)s	25%
18	S+C	geometry maintenance	25%
19	S+C	loss of detection	25%

A further analysis of the results ranks each problem and its underlying cause solely on the basis of cost. The top ten problems and their causes are as follows ([24][24.2]):

Table 20 The ten most important track problems and their underlying causes

Nr.	Component	Problems	Causes
1	Rail:	cracks and fatigue	creep forces
2	Rail:	cracks and fatigue	bad wheel/rail interface
3	Track:	bad track geometry	soft substructure/bad drainage
4	Switches and Crossings (S+C):	wear in switches	Substructure
5	Rail:	corrugations	vehicle/track interaction

Table 20 (*continued*)

6	S+C:	cracked manganese diamond (crossing)s	weld quality
7	S+C:	geometry maintenance	optimal maintenance regime?
8	Substructure:	unstable	soft substructure/wet track bed
9	Track:	bad track geometry	sub-optimal maintenance
10	Track:	bad track geometry	wrong/unknown stress free temperature

An analysis of the majority of the problems has been presented in the relevant paragraphs. In the following paragraphs we will analyse the causes for

- cracked manganese diamond (crossing)s
- bad track geometry
- weld quality

22.2 Switches and Crossings: Cracked Manganese Diamond (Crossing)

The failure of manganese diamonds (crossings) is analogous to wear in switches and therefore these priorities are equally relevant for cracked diamond (crossing)s.

A switch forms a discontinuity in any track where one is installed. It is a discontinuity with regards to track support due to the altered sleeper arrangement and during tamping operations it may require a separate operation or manual correction. A switch also forms a discontinuity for the wheel rail contact patch that may give rise to high transient vertical and creep forces (see 17.6.3.3). The high lateral accelerations to any vehicle not travelling in the straight ahead position will also cause high forces. The installation and maintenance of a switch is critical to its performance and any error may not be immediately apparent. For all these reasons, a switch will experience higher forces than plain line and the life of the switch will generally be reduced by plastic deformations, wear and/or fatigue cracks ([24][24.3])

22.3 Track: Bad Track Geometry – Wrong or Unknown Stress Free Temperature (SFT)

High rail temperatures result in the necessity to impose speed restrictions, and hence train delays, due to the increased risk of track buckling. This may be a risk due to the rail temperature exceeding the safe limit established from the rail stressing procedure, or a risk due to lack of confidence in the stress free temperature (SFT) in a track section. The problem for the track maintainer is that the SFT is almost always unknown and non-destructive or non-invasive measurement technology capable of continuously characterising the SFT of CWR does not exist. Climate change may result in more delays due to speed restrictions imposed as a result of hot rails, raising the priority for solutions to this problem condition ([24][24.3]).

22.4 Rail Welds

Rail welds are track discontinuities in regard to their metallurgical and mechanical properties. Rail welds tend to have lower hardness in the adjacent heat affected zones in comparison to the parent rail. Weld microstructures may also vary significantly from the parent rail. Two weld types comprise the majority of welds used in rail service, aluminothermic (thermite) welds and electric flash-butt (EFB) welds.

Thermite welds are by nature inclined to porosity and cleanliness problems, such as inclusions, typical of cast structures. Thermite weld production is highly operator dependant and as a result may experience significant variation in weld quality. EFB welds are much less operator dependant and produce the highest quality welds but are correspondingly more expensive and are less portable. Table 21 summarizes the types of failures associated with each weld process ([24][24.3]).

Table 21 Types of failures associated with each weld process

Typical Weld Degradation and Failure Modes	Aluminothermic Weld	Electric Flash-butt Welds
Railhead Batter / Dipping	Rail end misalignment. Mechanical Properties: • Hardness • Strength Distribution	Rail end misalignment Mechanical Properties: • Hardness • Strength Distribution
Railhead Shelling / Chipping (Surface and Subsurface RCF)	Mechanical Properties. Cleanliness.	Mechanical Properties. Electrode burns.

Table 21 (*continued*)

Horizontal Web Fractures	Fatigue initiation sites • Inclusions • Porosity • Mechanical damage • Weld geometry Wheel loading conditions • Vertical loading location • Lateral loading	Fatigue Initiation Sites • Inclusions • Mechanical Damage • Weld geometry (shear ridges) Wheel loading conditions Residual stress distribution • Vertical loading location • Lateral loading Residual stress
Vertical Base Fractures	Fatigue initiation sites • Inclusions • Porosity • Mechanical damage • Weld geometry Wheel loading conditions. Track support conditions. Residual stress distribution. Rail end misalignment. Thermal stress.	Fatigue initiation sites • Inclusions • Mechanical damage • Weld geometry (shear ridges) Vertical wheel loading. Track support conditions. Residual stress distribution. Rail end misalignment. Thermal stress.

23 Track Drainage

23.1 Introduction

There are four principal sources of water to consider that can affect the track [78]:

1. Rainfall directly on the track structure.
2. Surface water flowing toward and infiltrating the track structure.
3. Water flowing within the track structure, e.g., within ballast pockets (par. 19.6.2) or fill used to construct the embankment.
4. Groundwater.

Collecting and diverting surface water away from the track structure may be the most cost-effective portion of any subgrade maintenance program. The more water that is intercepted and diverted, the less water is available to infiltrate and potentially deteriorate the track structure.

Suitable track drainage is important to prevent the track formation from becoming saturated, that can lead to weakening and subsequent failure of the formation. So, adequate drainage must be installed in new or upgraded track, and existing drainage must be maintained so that it works effectively.

Track drainage consists of two types:

- Surface drainage
- Subsurface drainage.

In this chapter, prevention and correction of situations that may result in soft subgrades or embankment damage will be discussed. In more details, we will discuss:

- Common causes of subgrade problems and their recognition
- Remedies that may be used in typical soft subgrade or embankment instability situations

23.2 Surface Drainage

Surface drainage removes surface runoff before it enters the track structure, as well as collecting water percolating out of the track structure [78].

Basic grading of the ground on either side of the track is a form of surface drainage, and allows water flowing out of the track structure to be removed.

Shoulder grading may be used in very flat areas where it is difficult to get sufficient fall for either surface or subsurface drains.

K. Tzanakakis: *The Railway Track and Its Long Term Behaviour*, STTT 2, pp. 157–174.
DOI: 10.1007/978-3-642-36051-0_23

There are three main types of surface drainage. These are (Fig. 94):

- Cess drains (par. 23.2.1)
- Catch drains (par. 23.2.2)
- Mitre drains (par. 23.2.3).

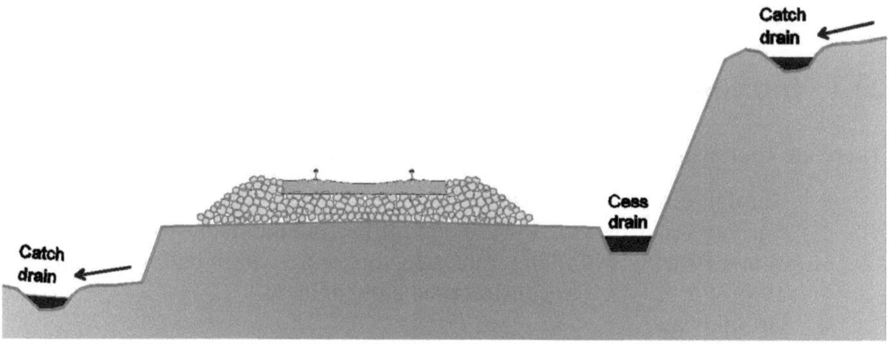

Fig. 94 Surface Drainage Types

> **Surface Drainage: Do's and Do Not's** [78]**:**
>
> - DO divert water away from track
> - DO keep cess drains clean and graded
> - DO NOT let water pond
> - DO NOT let water infiltrate the embankment

23.2.1 Cess Drains

Cess drains are surface drains located at subgrade (formation) level at the side of tracks, to remove water that has infiltrated through the ballast and is flowing along the capping layer towards the outside of the track formation. Cess drains are primarily intended for the protection of the subgrade (formation) by keeping the subgrade (formation) dry [78].

Cess drains are most frequently found in cuttings where water running off the formation cannot freely drain away.

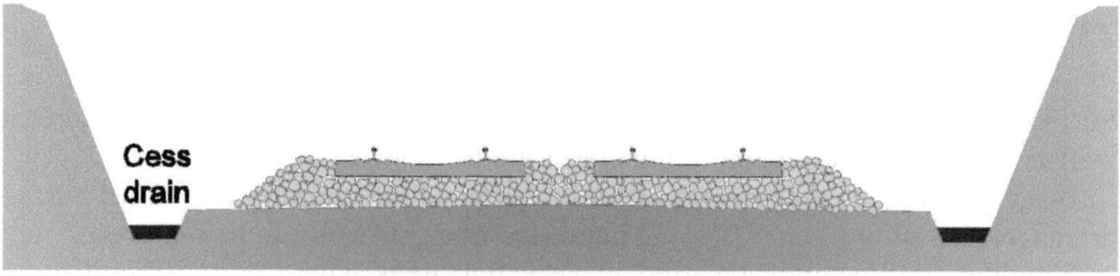

Fig. 95 Cess drain - Typical location

23.2.2 Catch Drains

The purpose of catch drains (also known as top drains) is to intercept overland flow or runoff before it reaches the track. They reduce the possibility of causing damage to the track or related structures, such as cuttings or embankments [78].

Catch drains are generally located on the uphill side of a cutting to catch water flowing down the hill and remove it prior to reaching the cutting (Fig. 96).

If this water was allowed to flow over the cutting face, it may cause excessive erosion and subsequent silting up of cess drains.

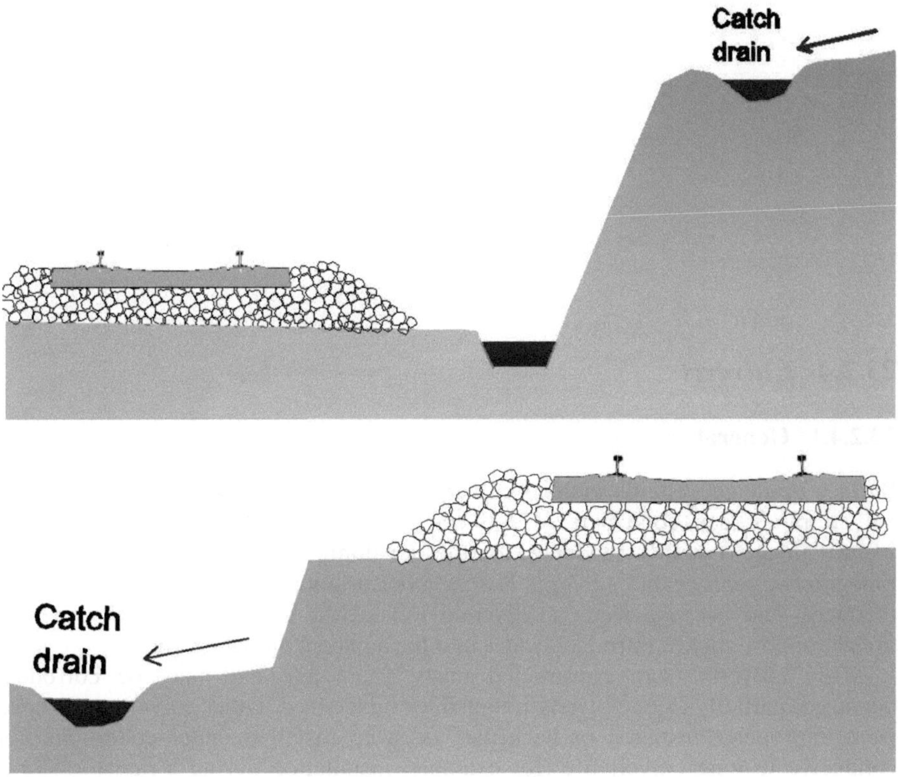

Fig. 96 Typical catch drains

Catch drains may be used alongside tracks that cut across a slight downhill grade.

23.2.3 Mitre Drains

Mitre drains (also called spoon drains or offshoot drains) are connected to cess and catch drains to provide an escape for water from these drains (Fig. 97). Mitre drains should be provided at regular intervals to remove water before it slows down and starts to deposit any sediment that it may be carrying [78].

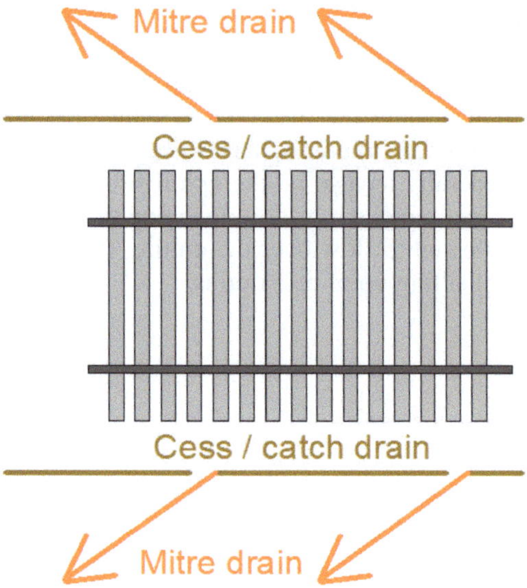

Fig. 97 Mitre drains

23.2.4 Culverts

23.2.4.1 General

Surface drainage systems include culverts. Improperly maintained or inadequate culvert pipes, can contribute to soft track situations. Also, culverts are often located in areas with a tendency for the development of soft track because of topography, *geology* and *geologic history*, poor original embankment construction practices, and because water is diverted toward the culverts – which provides greater opportunity to introduce water into the adjacent embankment.

Many culverts were constructed many years ago and may be corroded, damaged, partially or completely plugged, or undersized. Other culverts may have been improperly installed or backfilled or were installed under conditions not conducive to long culvert life (for example, installation during a rainstorm or a short track window) [86].

23.2.4.2 Culvert Maintenance

Leaking culverts (i.e. due to corrosion, poor installation etc.) may permit water to flow into the embankment soils. This water may then contribute to destabilization of the fill (see 23.5).

Culvert maintenance programs should include [86]:

- Replacement or repair of undersized and damaged culverts.
- Cleaning completely or partially plugged culverts.

- Relining of corroded pipes. Many methods for doing so are available. A qualified engineer should be consulted to evaluate the alternatives and develop recommendations.
- Replacing severely damaged undersized pipes, and pipes with vertical sags. If the existing damaged pipe is left in place in the embankment and a new pipe installed nearby, the inlet to the existing pipe should be plugged so that water does not continue to flow down it and leak into the embankment. Completely filling the abandoned culvert with soil, grout, or concrete should be considered if there is a potential for collapse of the culvert and impacts to the *track structure*.
- Installing internal couplers to repair or re-couple corrugated metal pipe joints that have partially pulled apart.

When a culvert joint or the culvert pipe itself has pulled apart, it should warn you that a more serious problem may exist at the site and something other than simply replacing or repairing the culvert may be required. Shifting and settlement of the embankment, progressive failure of the embankment over time, or relatively rapid slope movements (some of which may have happened in the past and the embankment "repaired" at that time) can result in culvert joints pulling apart or stretching and failure of the pipe itself.

Often this occurs i.e. in chronic soft track locations, in areas inclined to embankment problems. An embankment already inclined to instability may experience a further decrease in stability by water introduced into the embankment from the damaged culvert.

23.2.5 Maintenance Practices for Surface Drainage

Recommended maintenance practices include [86]:

- Improve surface drainage – make sure runoff runs off and does not pond.
- Divert surface water prior to it reaching the problem area.
- Prevent surface water from ponding and infiltrating the embankment.
- Deepen cess drains to 300 millimetres below the bottom of top subgrade (formation) (do not deepen cess drains to the point that the track section is undercut).
- Grade cess drain bottoms to improve runoff.
- Clean cess drains and culverts.
- Check culverts and ensure they are in good condition, have not rusted through, and have not pulled apart.
- Replace culverts that are too small with culverts of adequate capacity.
- Maintain culvert inlets and install systems to reduce the potential for the inlet to plug or for debris to wash into the culvert and plug it.
- Clean embankment cesses of undercutter, ballast cleaner, and cess drain cleaning debris. Grade cesses to drain water over side of the embankment without concentrating the flow (Fig. 106). Erosion protection measures may be required on the slope of the embankment.

23.3 Subsurface Drainage

23.3.1 General

Subsurface drainage is necessary for maintaining the integrity of the track subgrade (formation) and ensuring the stability of earth slopes [78].

Subsurface drainage is used for:

- drainage of the track structure
- controlling of ground water
- the draining of slopes.

Subsurface drainage shall be provided in locations where the water table is at or near earthworks level.

Subsurface drainage shall be provided along the cess, between, across, or under tracks as required. Advice should be sought from the Geotechnical Engineer before designing and installing sub-surface drainage.

Subsurface drainage systems shall be designed to take surface runoff, ground water and seepage, and water collected from other drainage systems to which the new system is being connected.

If a drainage system is required to remove ground water and seepage, a detailed hydrological and geotechnical investigation is required to determine the volume of water for the sizing of drains. The volume of water from other systems is determined from the outlet capacity of that system.

Subsurface drains are used where adequate surface drainage cannot be provided due to some restriction or lack of available fall due to outlet restrictions. Locations where these circumstances may occur are:

- Platforms
- Cuttings
- Junctions
- Multiple tracks
- Bridges

Water in the soil and in cracks that have opened in the ground can destabilize embankments or decrease soil strength. This water is often a major contributor to soft track conditions. Draining this water from the track section and embankment is fundamental to improving track and embankment performance.

Typical places within an embankment where water may be expected to accumulate or become trapped, and from which it should be drained, include [78]:

- Ballast pockets (par. 19.6.2)
- Cracks in embankments
- At the contact between the subgrade and a relatively rigid slab that overlies the subgrade but is below the ballast.
- At the contact of more permeable rock or soil with less permeable materials.

A method commonly employed to drain water from track sections and embankments is to construct gravel-filled trench drains (Fig. 98 to Fig. 100).

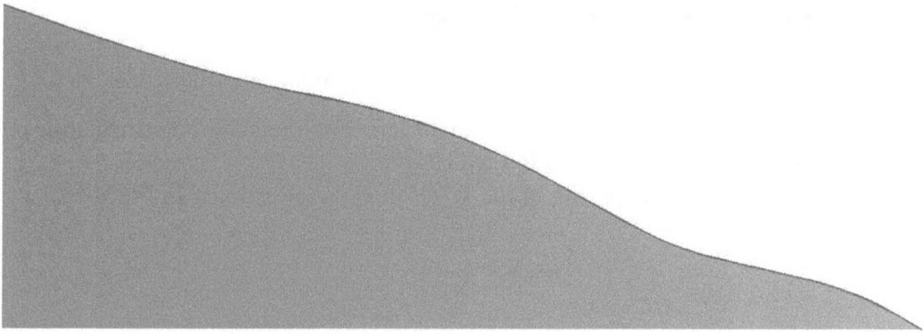

Fig. 98 A slope before the construction of a track

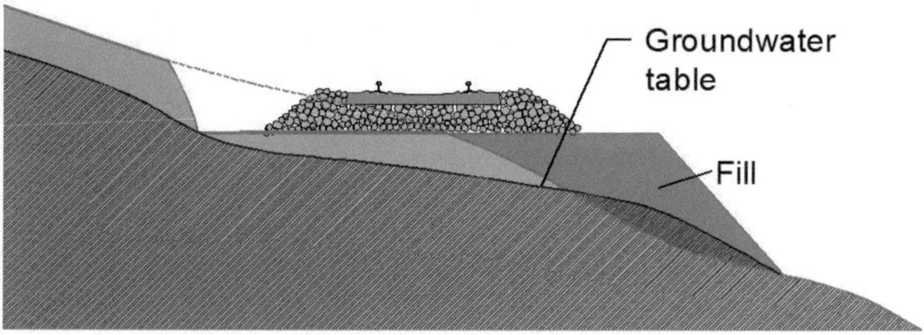

Fig. 99 Groundwater elevation in fill before installing trench drain

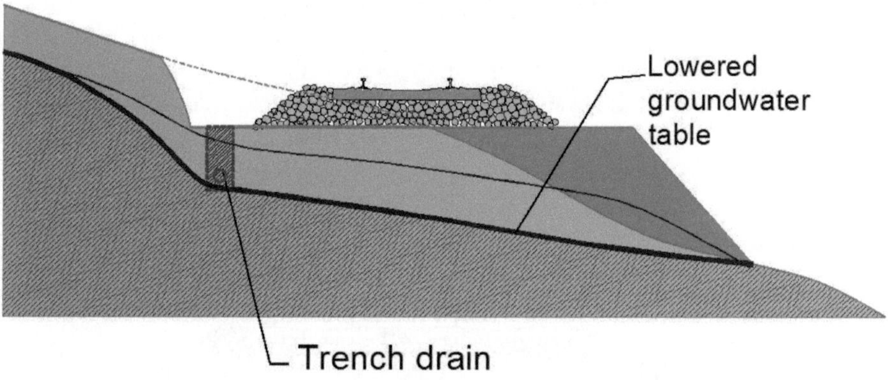

Fig. 100 High groundwater condition causing embankment instability. Lowered groundwater condition resulting from trench drain installation also depicted (after [86])

23.3.2 *Functions of Subsurface Drains*

Subsurface drainage systems perform the following main functions [78]:

- Collection of infiltration water that seeps into the subgrade (formation) or capping layer), as shown in Fig. 101.
- Draw-down or lowering of the water table, as illustrated in Fig. 102.

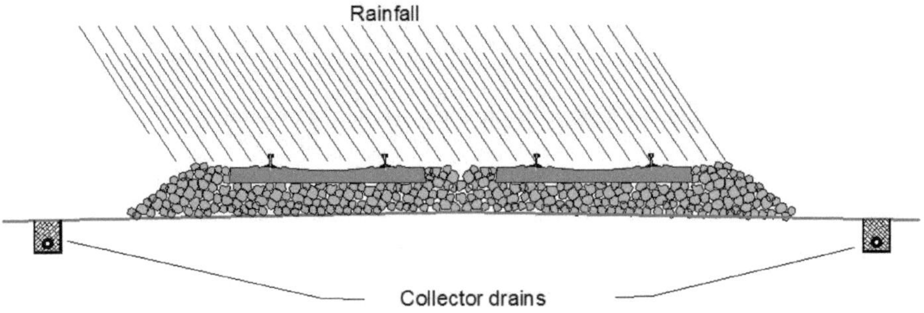

Fig. 101 Collection of water seeping into the ballast structure

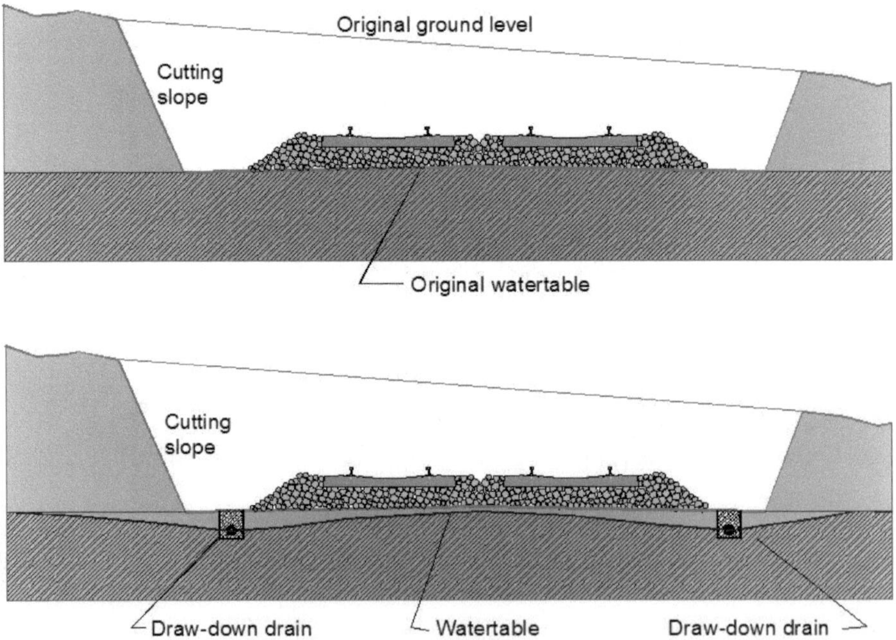

Fig. 102 Lowering the water table

Subsurface Drainage Systems are effective for

- draining water from soil
- draining water from ballast pockets
- draining water from cracks in the ground
- lowering the groundwater elevation
- intercepting water flowing toward the track

Subsurface Drainage: Dos and Do Not's

- DO drain ballast pockets
- DO NOT construct deep drains without the assistance of experienced personnel

23.4 Indicators of Poor Drainage

(After [85])

Wet Ground near or on the Embankment	
Ponded Water Adjacent to Track	

Cattails[39] on Slope above Track 	
Water ponded in ditch and wetlands[40] plants	

23.5 Failure Cases

23.5.1 Track Geometry Faults due to Poor Drainage

Fig. 103, shows a "perfect" track, but with localized track geometry faults, measured at a certain track position by the Austrian Measurement Coach EM 250, and at different time periods (after Prof. P. Veit, Technical University of Graz).

[39] *Typha* is a genus of about eleven species of monocotyledonous flowering plants in the family Typhaceae. The genus has a largely Northern Hemisphere distribution, but is essentially cosmopolitan, being found in a variety of wetland habitats. These plants are known in British English as bulrush, bullrush, or reedmace, in American English as cattail, catninetail, punks, or corndog grass, in Australia as cumbungi or bulrush, and in New Zealand as raupo.(source: wikipedia).

[40] A wetland is an area of land whose soil is saturated with water either permanently or seasonally.

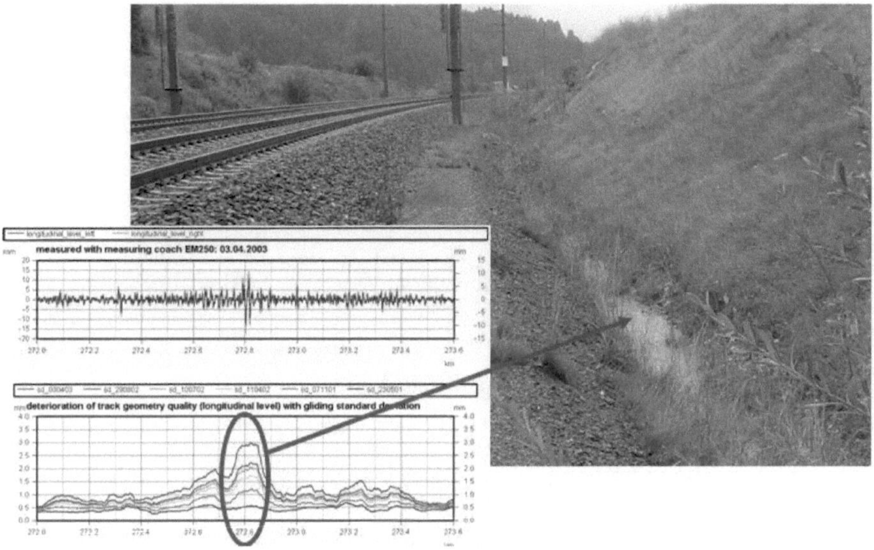

Fig. 103 Track geometry faults due to poor drainage[41]

The ponded water infiltrates in the ground, with a local variation of the track stiffness, leading to a track softening at that location (Fig. 104).

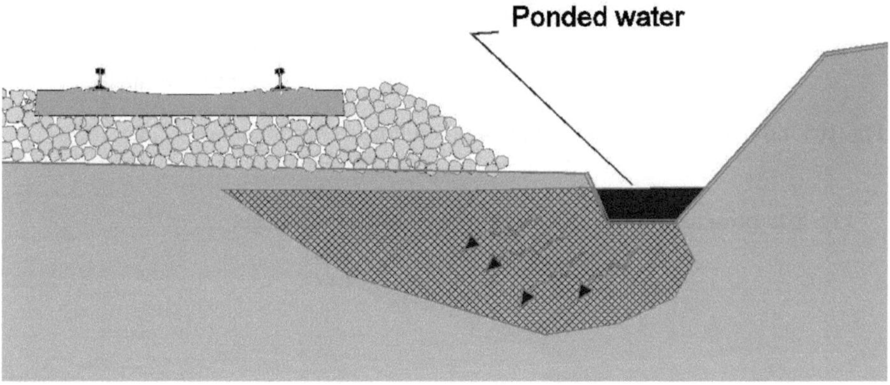

Fig. 104 Ponded water

23.5.2 Failure due to Poor Drainage: Infiltration and Embankment Failure

In Fig. 105 an infiltration and embankment failure due to poor drainage is presented.

---

[41] Source: Dr. Florian Auer, ÖBB".

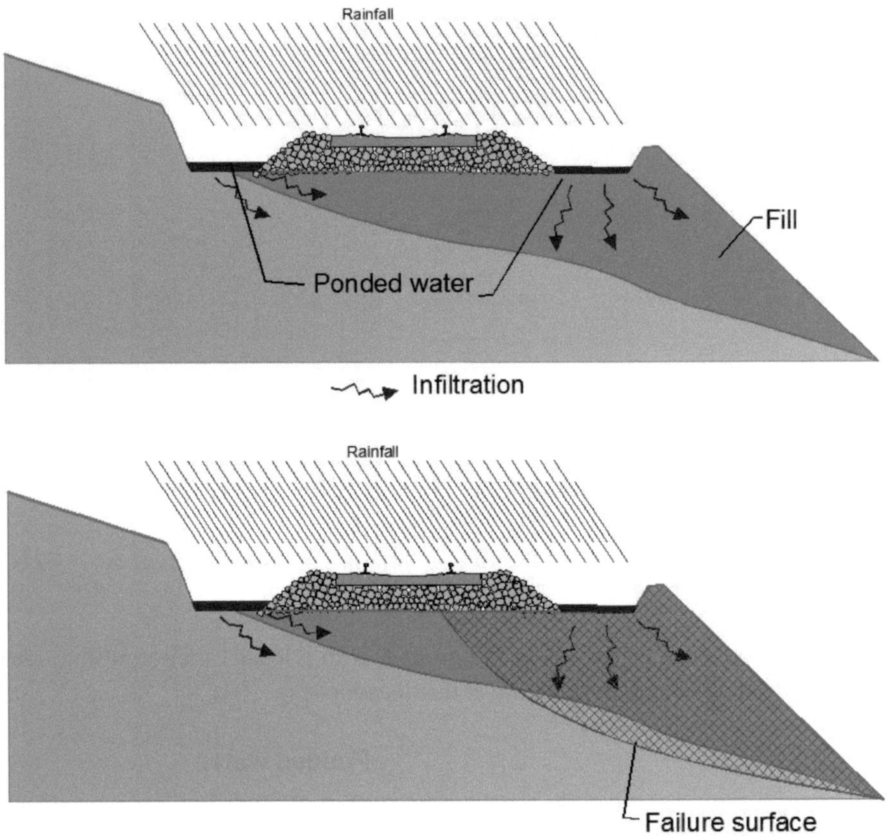

Fig. 105 Infiltration and embankment failure resulting from poor drainage (after [85])

Fig. 106 presents remedies for improved surface drainage.

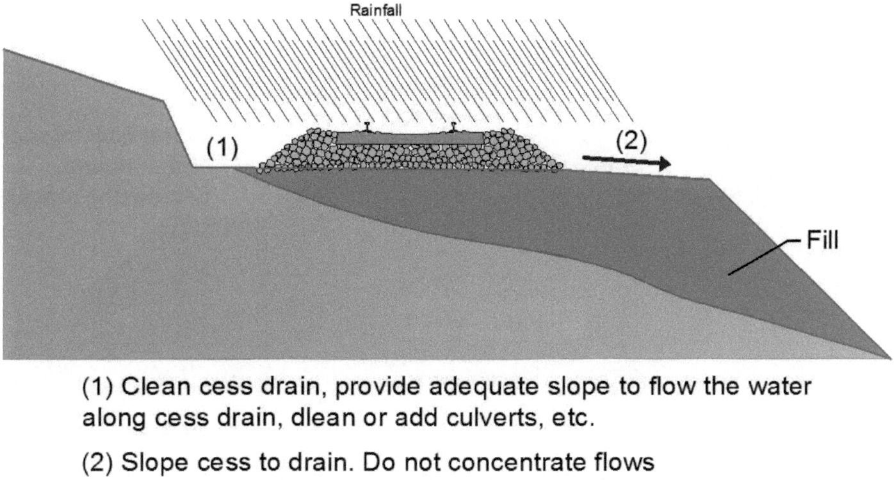

Fig. 106 Improved Surface Drainage [85]

23.5.3 Embankment Failure Resulting from Impounded Water

Embankment failure resulting from impounded Water is possible when no culvert is installed, or the culvert is plugged, so that the impounded water cannot flow away. The water flows through the embankment, decreasing stability which can lead to a failure surface (Fig. 108 to Fig. 110).

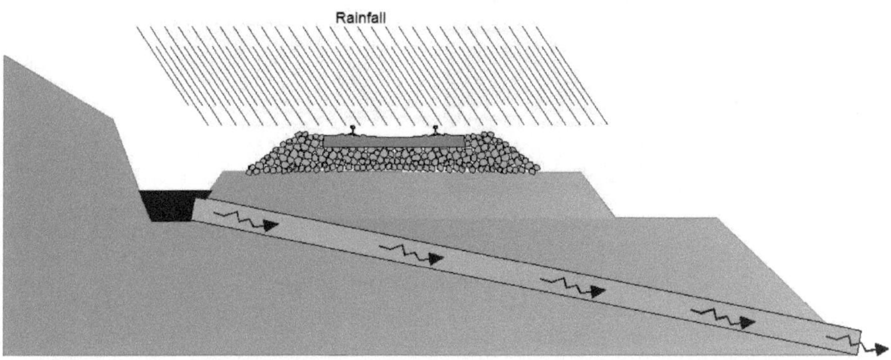

Fig. 107 A good functioning culvert

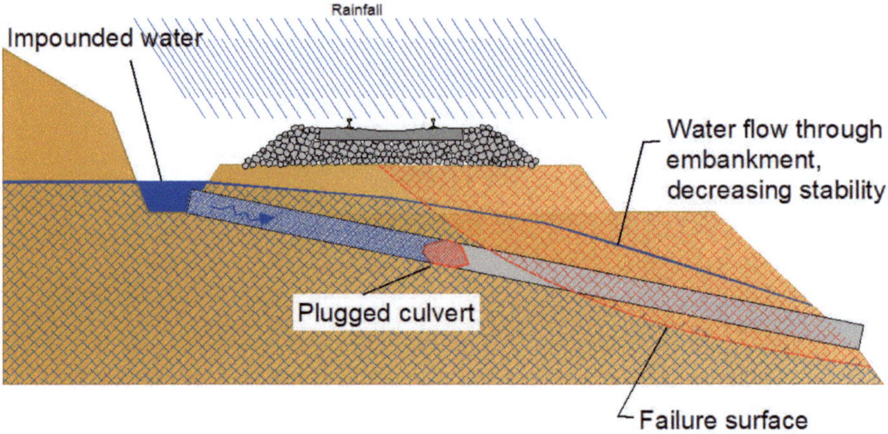

Fig. 108 Embankment Failure Resulting from Impounded Water [86]

Fig. 109 Plugged culvert [86]

Fig. 110 Washout because of plugged culvert [86]

23.5.4 Erosion of Embankment Slope by Water Flowing over the Embankment

In this case, the culvert is plugged or no culvert installed, the impounded water is in the level of the rail/sleeper. Water flow over or through the track, eroding embankment (Fig. 111, Fig. 112).

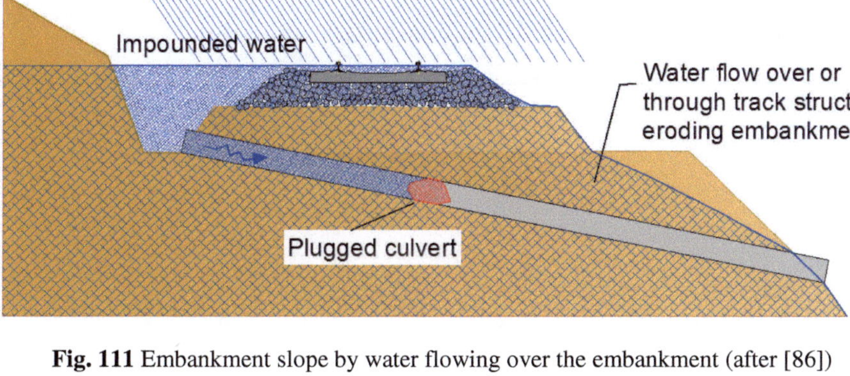

Fig. 111 Embankment slope by water flowing over the embankment (after [86])

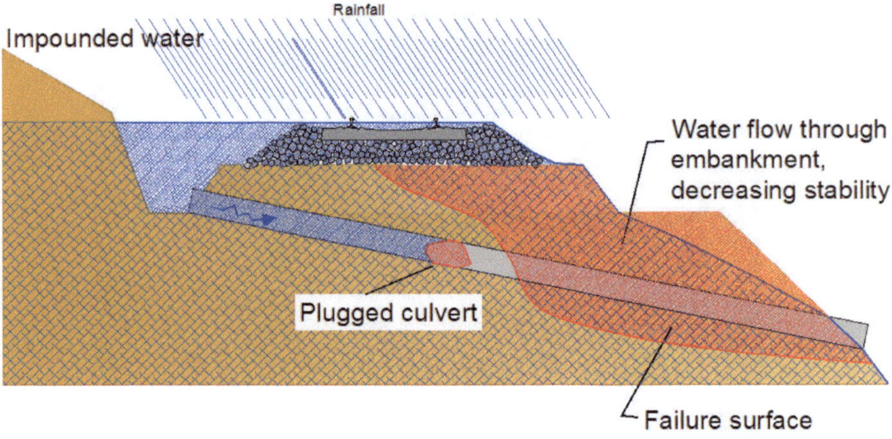

Fig. 112 Erosion of Embankment Slope by Water Flowing Over the Embankment (after [86])

23.6 Keep in Mind ...

1. Water is a major contributor to soft track ... and the key to reducing the occurrence of soft track is ... ...**DRAINAGE**
2. ... Water trapped in a ballast pocket is a major contributor to many soft track situations ...
3. Proper surface drainage is probably the least expensive and most easily implemented measure for preventing soft track problems or improving performance of soft track areas ...

   [86])

23.7 Bad Examples

	Cess drain plugged by old ballast.
	Cess drains plugged by old ballast hindering the water to flow to the cess drain.
	Renovation of a track in a railway station. The track is laid in a "vessel". Water cannot be drained.

	Electrification pole and cable channels in the cess drains, hindering the water flow (drainage). Danger of pounded water.

24 The Track Deterioration Model

It is well known that deterioration depends on the present quality level:

- good track behaves well (deteriorates more slowly),
- poor one deteriorates faster

This general observation means that a track behaves according to its quality. Mathematically it can be expressed as ([74], [75]):

$$Q = Q_o \cdot e^{b \cdot t}$$

Where:

Q the actual track quality
Q_0 the initial quality
b the deterioration rate coefficient
t the time

Track quality is deteriorating due to traffic, showing various faults. The development of the track quality is shown in Figure 113:

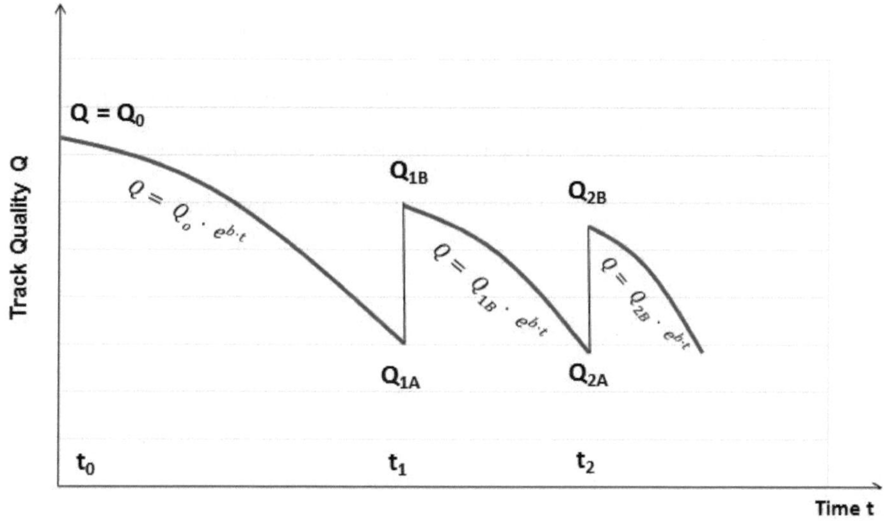

Fig. 113 The development of the track quality

K. Tzanakakis: *The Railway Track and Its Long Term Behaviour*, STTT 2, pp. 175–176.
DOI: 10.1007/978-3-642-36051-0_24

The next table describes the development of the track quality during its life in relation to maintenance interventions at certain times, in order to keep the track quality/functionality at the desired level.

Time	Event	Quality level
Time t_0	Construction or rehabilitation of the track. The track has its initial (best) track quality Q_o.	Q_0
Time t_1	Gradually track degradation. Interventions are necessary in order to keep the track quality/functionality at the desired level.	Q_{1A}
	Interventions improve the track quality at quality level Q_{1B}. The new track quality doesn't reach the initial quality ($Q_{1B} < Q_0$).	Q_{1B}
Time $t_{2,} \dots t_n$	The track degradation cycles continuous. The track quality after every intervention never reaches the previous quality level ($Q_{nB} < Q_{(n-1)B}$ <….< Q_{2B}<Q_{1B}<Q_0).	$Q_{2B} \dots Q_{nB}$

From the previous examination, we can draw the following conclusions / positions ([74][75]):

1. Description of quality must consist of status data and their behaviour over time.
2. Track quality isn't a stable situation (without changes), track quality is a permanent procedure.
3. The development of the track quality geometry is predictable.

In the next paragraphs we will extensively analyse those positions.

Section V: Track Quality

> **"Quality is not an act, it is a habit."**
> Aristotle
>
> **"Quality is never an accident; it is always the result of high intention, sincere effort, intelligent direction and skilful execution; it represents the wise choice of many alternatives."**
> William A. Foster

Quality should be the final goal for the Infrastructure Manager. In this section we examine the importance of good quality and how quality affects the long term behaviour of the railway track.

25 The System "Railway Infrastructure"

Key Terms: *Track quality behaviour, quality and safety, track quality and dynamic load*

25.1 Introduction

The system "railway infrastructure" can be described by three basic constituents: the capacity of the network, the substance and the quality of infrastructure (Figure 114) [77]:

- **Capacity** (maximal number of operable and marketable train-paths in a given time span, on a given part of the network),
- **Substance** (average remaining life time of infrastructure components) and
- **Quality** (quality of track's geometry and of its components).

Capacity, substance and quality are **interrelated parameters** and **must jointly be adjusted**.

Capacity, substance and quality may be represented trough the following figure.

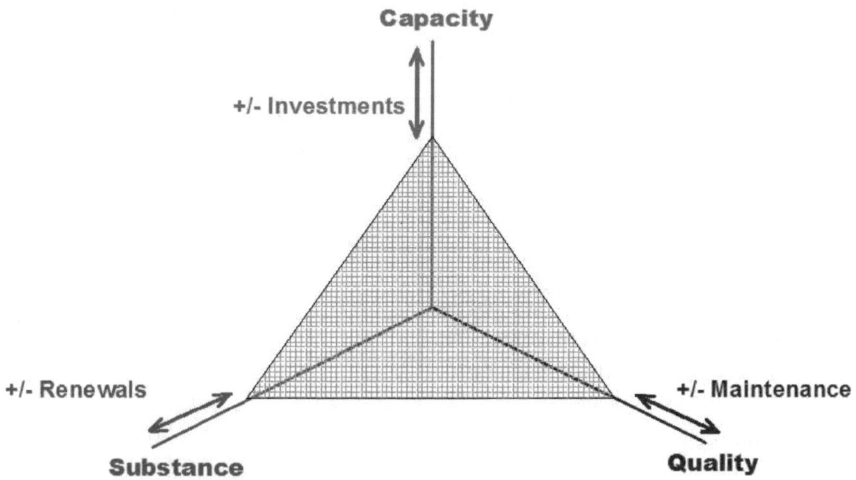

Fig. 114 System "Railway infrastructure"

Managing infrastructure amounts to the same as setting the three parameters at the most appropriate level, in order to maximize the efficiency of the whole system:

K. Tzanakakis: *The Railway Track and Its Long Term Behaviour*, STTT 2, pp. 179–183.
DOI: 10.1007/978-3-642-36051-0_25

- capacity can be adjusted through adapting investment policies
- infrastructure substance can be tuned through a suitable renewal policy and
- quality can be adjusted with the maintenance policy.

Capacity, substance and quality of a network are tightly related together, as can be seen in a case of loss of the rail infrastructure due to insufficient budgets, Figure 115.

25.2 Case of Loss of the Rail Infrastructure due to Insufficient Budgets

<u>**Initial situation – YEAR 0**</u> Infrastructure is in a good state. The quality is high enough to guarantee the expected level of service quality.	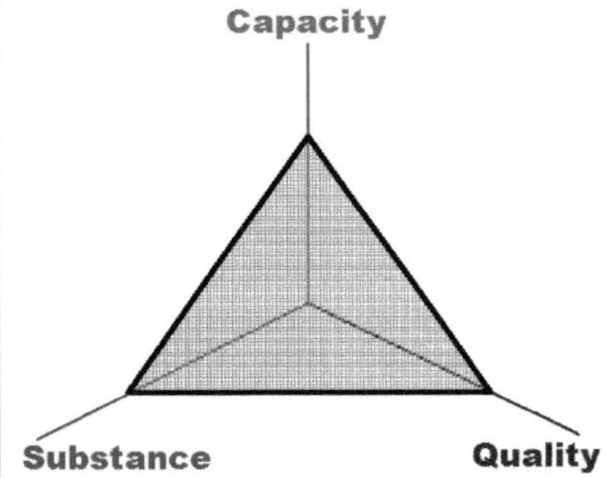
<u>**First step – YEAR +1 / +2**</u> Limitation of budget for renewal actions or decrease of available intervals for renewal. The consequences of such a policy are not yet visible as the infrastructure has enough substance. More capacity is available.	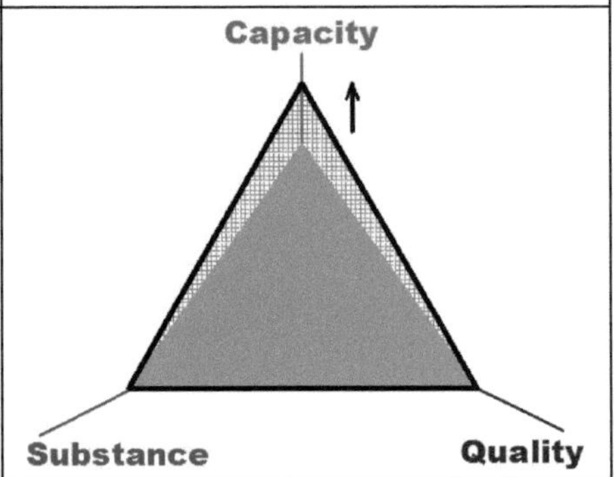
<u>**Second step – YEAR +3 / +4**</u> In order to keep quality standards, more maintenance has to be done. The infrastructure grows older and needs more maintenance actions. The availability of capacity decreases too.	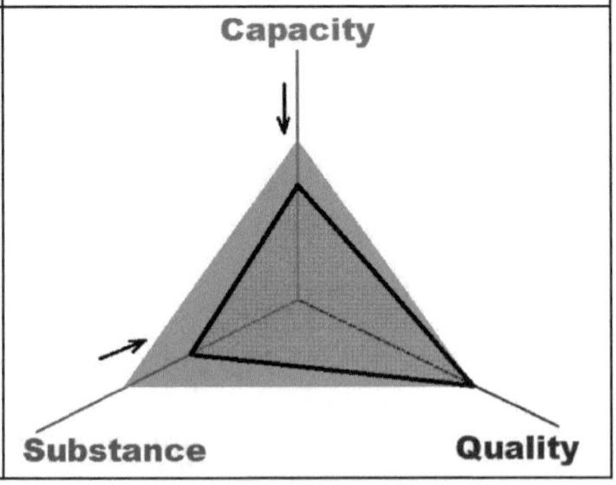

Fig. 115 Loss of the rail infrastructure due to insufficient budgets (after [77])

Third step – YEAR +6 / +8 It becomes hard to keep quality standards as needs in maintenance grow very quickly. There aren't enough resources to achieve the maintenance. Speed restrictions have to be set in order to guarantee the safety of the traffic. There is a significant decrease of available capacity.	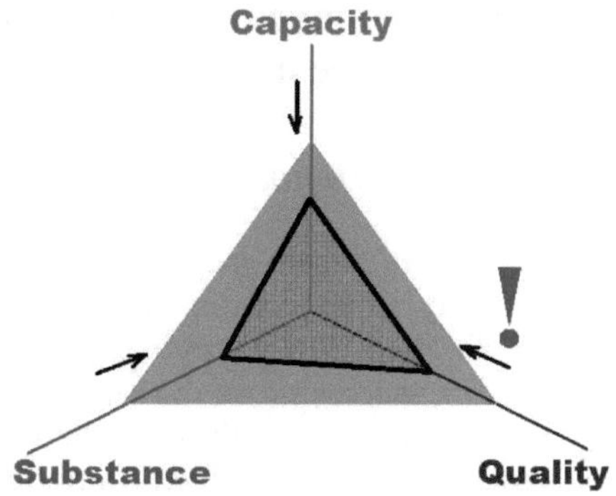
Final step – YEAR +10/+... The system "infrastructure" is totally unstable. Maintenance costs are out of control, the quality decreases. There is a significant lack of capacity. Massive renewal actions have to be launched in order to reverse the trend and to bring the network to its balance. These renewal actions require high investments and works lasts several years, years during which railways operators experience serious difficulties to offer an acceptable service.	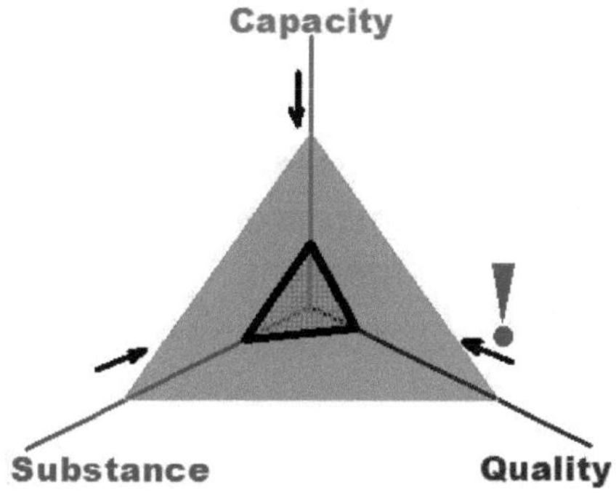

Fig. 115 (*continued*)

The result of an insufficient budget is a dangerous loss in substance that unfortunately may become visible only in mid- or long-term range; the infrastructure lives on its substance. Considering the fact that after a certain age it is impossible to guarantee the quality of the infrastructure by a maintenance action, a renewal action has to be undertaken. An infrastructure living on its substance means that the average age of the infrastructure is constantly growing, leading to a point where an important part of the network should be renewed at the same time. This produces budget problems, machines and manpower bottlenecks and, usually, as renewal actions can't be done quickly enough, speed and traffic restrictions must be enforced. The system "rail" is then unstable.

This example gives evidence that certain basic decisions on the maintenance and renewal policy as the ratio between maintenance and renewal budgets have a decisive influence on the long-term network capacity. **It is therefore crucial for any Infrastructure Manager to adapt a long-term M&R policy, integrating the evolution of capacity as well as maintenance and renewal.**

25.3 Dependence between Railway Infrastructure Capacity, Substance and Quality

25.3.1 First Dependence: Capacity vs. Maintenance and Renewal

An increase of traffic induces an escalation of infrastructure deterioration and consequently an expansion of needs in its maintenance and renewal (hereafter M&R). Related M&R engineering works induce then an increase of track possessions, worsening so the saturation rate of the network. If engineering works are not undertaken, condition of the infrastructure will sooner or later deteriorate and generate permanent speed restrictions (to avoid a decrease of the safety level). The figure below illustrates the problem [77].

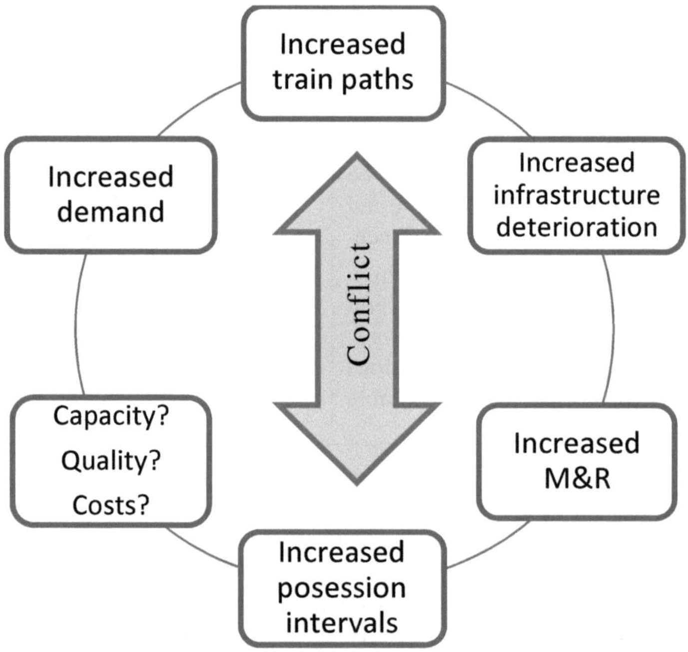

Fig. 116 Technical interrelation between capacity and M&R

The dependence between capacity and M&R occurs not only on the technical side of infrastructure management but obviously also on the economical field:

- the marketing department's goal is to sell as many slots as possible to satisfy customer demand.
- The maintenance department's task on the other hand is to keep the infrastructure on a defined quality level. M&R works have to be carried out at the lowest possible costs, which imply the use of heavy machinery and large track possessions inherently consuming capacity, hereby reducing marketable train-paths.

A demand increase leads to a bigger train-paths requirement, which then results in a growing deterioration of infrastructure and consequently more intensive maintenance and renewal (M&R) works. In some cases new infrastructure is necessary to avoid bottlenecks (capacity investments). Both of the last points (maintenance and renewal works growth and capacity investments) induce an increase of infrastructure possession intervals for the works. More infrastructure possession intervals decrease the available trainpaths and the infrastructure may not be able to supply the market. This is a conflict with which the infrastructure management has to deal (Figure 116).

Thus, the infrastructure management includes two strongly linked aspects: the capacity management and the resource management.

- The **capacity management** aims at making train-paths available as required by the market, at an acceptable level of quality, taking into account the infrastructure possession time required by M&R works and an acceptable cost.
- The **resource management** aims at ensuring the required Reliability-Availability-Maintainability-Safety (RAMS) of the infrastructure, minimizing its life cycle cost.

25.3.2 Second Dependence: Quality vs. Substance

As already examined the most significant factor contributing to infrastructure deterioration by far is the dynamic load of trains running on the track. The dynamic load is directly related to the axle load and track geometry. The two elements are closely linked in the complex process of deterioration.

The bad state or quality of material contributes significantly to the deterioration of the track geometry. Bad track geometry produces higher dynamic loads on the material, which accelerates the degradation of the material, and thus significantly reducing the substance. A detailed analysis is given in Section IV.

26 What Is Quality?

26.1 Overview

There are general questions referring to track quality, which are dominant in periods with big budget constraints:

- Do we really need quality?
- How much quality is technically necessary?
- Is quality economically justified?

Quality is a result partly of investment and renewal, and on the other hand a result of maintenance [75]:

- The role of investment and re-investment is the creation of an adequate output quality.
- The task of the maintenance is to "translate" this into an appropriate service life of the track.

26.2 The Track Quality Behaviour during the Years

The quality of a section of track at the beginning of its life (after its construction) determines its future behaviour. From various investigations it is known that the degradation rate of a certain track section is like a typical fingerprint. The final

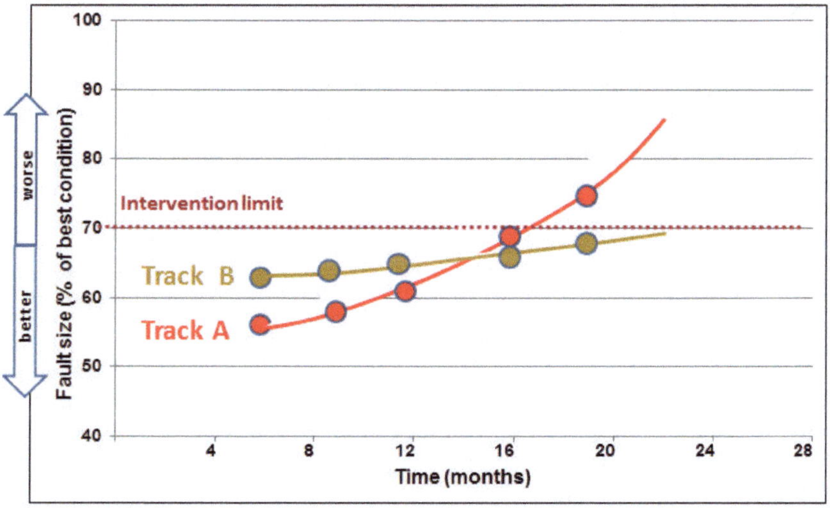

Fig. 117 The track behaviour of track in time

K. Tzanakakis: *The Railway Track and Its Long Term Behaviour*, STTT 2, pp. 185–190.
DOI: 10.1007/978-3-642-36051-0_26

quality of a track section at the beginning of its life determines its behaviour over time. The initial quality affects the deterioration rate reflecting the suitability of the track components for the traffic load [76].

As we can follow on Figure 117[42], the two tracks presented have a different behaviour in time. Although "track A" has a better initial quality, it deteriorates faster than "track B". ("track A" reaches the intervention limit in 16 months, and needs maintenance. For "track B" the intervention limit is reached in about 23 months).

26.3 The Importance of Good Initial Quality

We start with two examples, demonstrating the importance of good initial quality.

Example 1: Comparison of the initial quality of two tracks

In Figure 118, two new constructed tracks are compared in time: **Track 1** and **Track 2**, having initial track quality Q_{01} and Q_{02} (t_0 is the time after track construction, when they are given in operation).

In this example, Q_{01} is 90% of that of a perfect track. For Track 2, Q_{02} the respective value is 80% (that means that the two tracks have a difference of about 10% in quality). We suppose that the intervention limit is set to 70%.

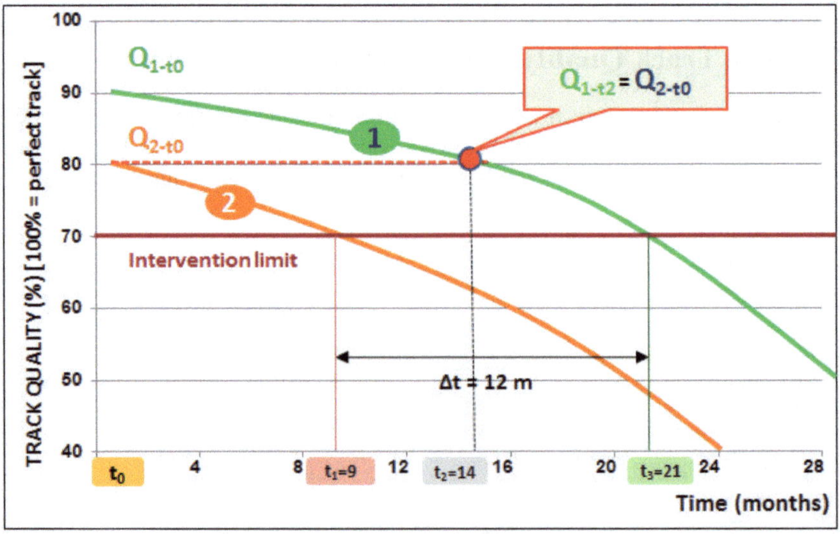

Fig. 118 The importance of good initial quality -1

Following this example, after t_2=14 months "Track 1" will have the same initial quality as "Track 2" ($Q_{1\text{-}t2} = Q_{2\text{-}t0}$). Track 1 will reach the intervention limit (an indication of starting maintenance needs) in t_3 = 21 months. The respective time for "Track 2" is t_1 = 9 months.

[42] After Prof. Peter Veit seminars (Technical University of Graz).

Conclusion:
Although the initial quality of two tracks doesn't differ so much, there is a significant difference in the starting of maintenance needs.

Example 2: Comparison of the standard deviation of the profile-faults of two tracks

In Figure 119 the standard deviation of the profile-faults of two tracks is compared according to traffic load. For Track 1 and Track 2, after their construction the profile-faults has been measured with the track recording car, and found for "Track 1" σ_0 = 0,8 mm and "Track 2" σ_0 = 1 mm (difference is 0,2 mm). We suppose that the intervention limit is σ = 2 mm.

"Track 1" will reach the intervention limit after about 18 million tonnes. The respective value for "Track 2" is about 9 million tonnes.

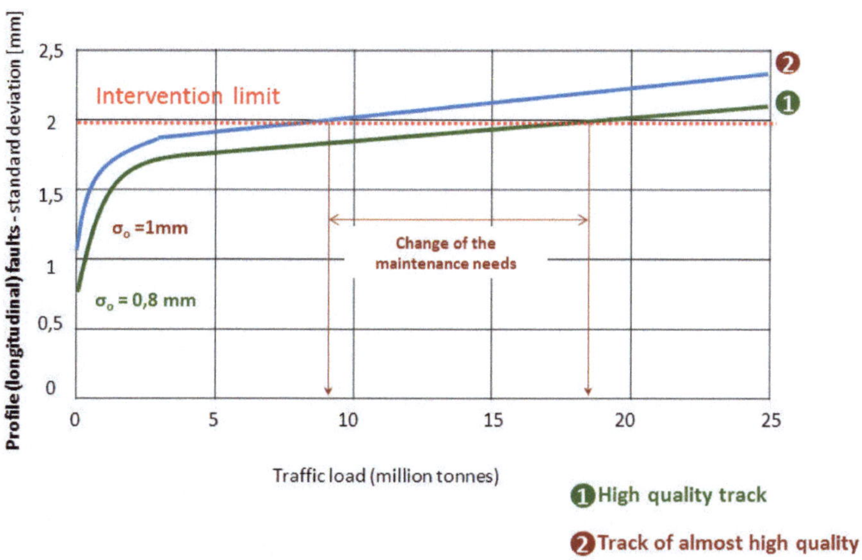

Fig. 119 The importance of good initial quality -2 [76]

Conclusion:
Although the track faults of two tracks are very similar at the beginning of their life, there is a significant difference in the starting of maintenance needs.

Conclusion drawn of the two examples

Figures above show the relationship between initial track quality and track deterioration rate. The worse the initial quality of a track, the more quickly deteriorates the track geometry.

The initial quality plays a substantial role for the life of the track. Already small variations in the initial quality significantly influences the maintenance needs.

There is a pattern of track behaviour [74]:

- good track behaves well,
- poor one deteriorates faster.

26.4 Quality and Safety

26.4.1 Introduction

In the figures below we can see a track in very bad condition[43] and a very good one. What track is safer?

(a1)

(a2)

(b)

Fig. 120 A low and a high quality track

Answering this question most would say: *"the track on the right"*, thinking that quality is directly connected to safety. The right answer is: *"both tracks can have the same level of quality"*. But there is a precondition that the trains are running with the proper speed, set by the track engineer, in respect to the track faults. For this speed both tracks are safe, having equal safety level. The track on the left

[43] At Montmorency on the Hurstbridge rail line (Australia).

could be very safe for 30 km/h and very unsafe for 60 km/h. The track on the right could be very safe for 160 km/h and very unsafe for 250 km/h.

Conclusion:
Quality has nothing to do with safety[44].

26.4.2 Safety and Maintenance

Due to train traffic, track geometry is deteriorating. The aim of the track engineers is to ensure the safe running of the trains at the designed performance level without speed restrictions, which decrease the performance level of the track.

Let's examine the development of a track fault using the graph below (Figure 121):

- Fault size and failure probability

  A very new track has (usually) very small sized faults, which are growing in time, as track is deteriorating. Until they reach a certain value (at level f_0) there is no probability of a failure due to track geometry (in the graph below section f_{00} – f_0 on the vertical axis). When the faults are growing beyond this level, there is a risk of failure (in the graph bellow section f_0 – f_{100}). If the track faults exceed a certain value (level f_{100}) the probability of a failure is 100%.

- Fault development and safety
  - At time t_0 the track is new and the faults are very small sized (youth period of the track, see 18.2).
  - At time t_1 a fault is found, with a size exceeding the *alert limit*. The fault is growing. Until the size f_0 is reached (at time t_3), the train can operate with full safety[45]. If the fault is not restored at t_3 (when reaching the *immediate action limit*), speed restrictions must be set (to keep the safe running of the trains).
  - Assuming that the Infrastructure Manager executes proper maintenance, he would start the maintenance to restore the fault at earlier stage (when reaching the *intervention limit*), that is at time t_2, in order to have a safety margin[46].

    The safety margin is defined in the maintenance strategy of the Infrastructure Manager and has not to do with safety but is set up only with economic criteria [74].

[44] Prof. Peter Veit, [74].
[45] Without any speed restrictions.
[46] In par. 32.9 and 32.6 we will discuss in a big extension the different limit levels (alert – intervention - immediate action limit) and the different maintenance strategies.

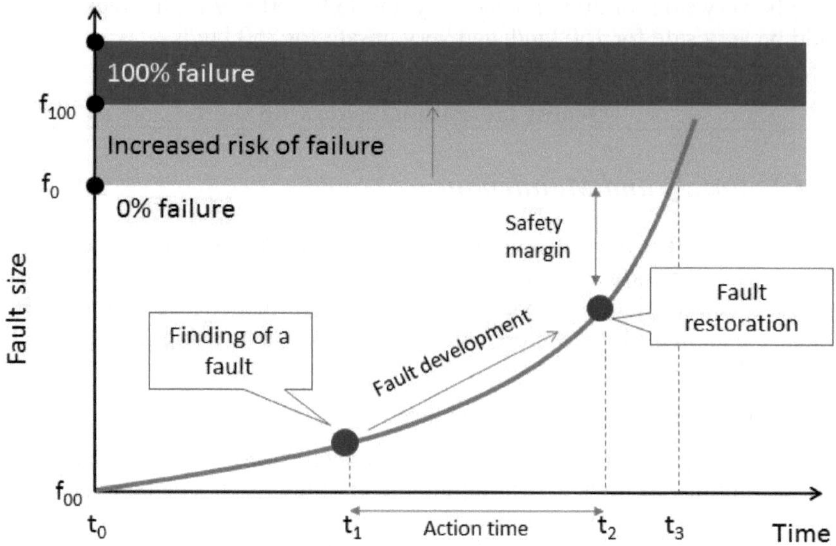

Fig. 121 The fault development and the safety

27 Factors Affecting the Track Quality Behaviour

27.1 General

As we examined in Section IV, "deterioration or degradation is the reduction of the original quality due to use and/or environmental influences". In par. 18.2 we identified that three main groups of factors may be distinguished that contribute to the deterioration of railway infrastructure:

- **Use:** wear by physical contact, static and dynamic load
- **Environment:** climatic influence, water
- **Failures:** faulty components, bad construction

In most of the cases it is not just one of these factors that causes deterioration, but a combination of them.

The following figure summarises the factors affecting the track degradation and thus its quality behaviour.

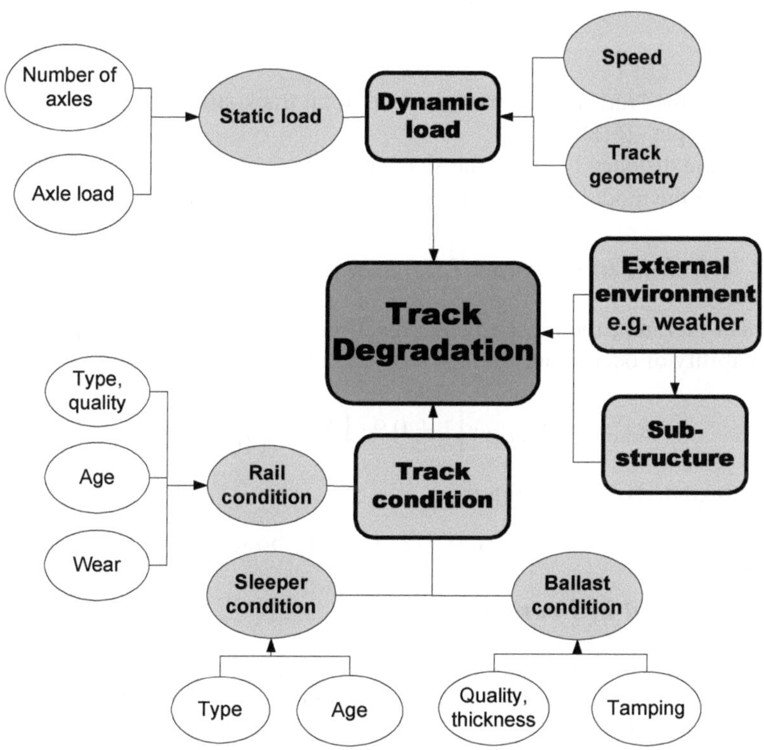

Fig. 122 Factors affecting the track quality behaviour

K. Tzanakakis: *The Railway Track and Its Long Term Behaviour*, STTT 2, pp. 191–194.
DOI: 10.1007/978-3-642-36051-0_27

27.2 The Interaction of Track Quality and Dynamic Load

Actions on track according to international literature [134]: the theoretical analysis is based mainly in Winkler's theory (see par. 15.3.2) of an infinite beam on elastic foundation. In international literature four methods are –mainly- cited.

27.2.1 Method in the American Literature

The most adverse reaction/action on each support point (sleeper) is given by (see [135]):

$$R_{max} = \bar{A}\text{stat} \cdot \left(1 + \frac{D_{33} \cdot V}{D_{wheel} \cdot 100}\right) \cdot Q_{wheel}$$

where:

- D_{33} in inches the diameter of a wheel of 33 inches,
- D_{wheel} in inches the wheel's diameter of the vehicle examined,
- V the speed in miles/hour, and
- $\bar{A}_{stat}$ is the same as in equations of the European literature below and it is given by:

$$\bar{A}_{stat} = \frac{1}{2\sqrt{2}} \cdot \sqrt[4]{\frac{l^3 \cdot k_{total}}{E \cdot J}}$$

where:

- k_{total} the "rail support modulus" or "total track stiffness (static)" in kN/mm (see par. 16.1),
- ℓ the distance between the sleepers,
- E, J the modulus of elasticity and the moment of inertia of the rail.

27.2.2 Method in German Literature

In German literature the most adverse reaction R_{max} per sleeper is dependent upon the probability of occurrence and for 99.7% probability is given by ([136],[137]):

$$R_{max} = Q_{wheel} \cdot \left[1 + 0{,}9 \cdot \left(1 + \frac{V - 60}{140}\right)\right] \cdot \bar{A}_{stat}$$

for $V \geq 60$ km/h. If $V < 60$km/h then

$$R_{max} = 1{,}9 \cdot Q_{wheel} \cdot \bar{A}_{stat}$$

27.2.3 Method in French Literature

There is also the method cited in French literature ([134],[138],[139]) covering a probability of occurrence 95,5% and distributing the total acting load with reaction per sleeper $1{,}35 \cdot \bar{A}_{stat} \cdot Q_{total}$ as follows:

$$R_{max} = \bar{\text{A}}_{\text{stat}} \cdot 1{,}35 \cdot \left[Q_{wheel} \cdot \left(1 + \frac{Q_\alpha}{Q_{wheel}}\right) + 2 \cdot \sqrt{\sigma(\Delta Q_{NSM})^2 + \sigma(\Delta Q_{SM})^2}\right]$$

where

- Q_{wheel} the static wheel load,
- Q_α the load due to cant deficiency,
- 2 coefficient of dynamic load for a 95.5 % probability of occurrence,
- $\sigma(\Delta R_{NSM})$ the standard deviation of the dynamic load due to non-suspended masses,
- $\sigma(\Delta R_{SM})$ the standard deviation of the dynamic load due to suspended masses.

27.2.4 Method in Greek Literature

After an -over 10 years- investigation program, in the Greek network, due to the appearance of extensive cracks in concrete ties laid on track, in a percentage over 60%, a method was developed that is able to predict the observed conditions on track ([140]). The actions on track panel are calculated through the following equation covering a probability of occurrence 99,7%:

$$R_{max} = (Q_{wheel} + Q_\alpha) \cdot \bar{\text{A}}_{dyn} + 3 \cdot \sqrt{\sigma(\Delta Q_{NSM})^2 + \sigma(\Delta Q_{SM})^2}$$

where

- Q_{wheel} the static wheel load,
- Q_α = the load due to cant deficiency,
- $\bar{\text{A}}_{dyn}$ = dynamic coefficient of sleeper's reaction,
- 3 coefficient of dynamic load for a 99,7 % probability of occurrence,
- $\sigma(\Delta R_{NSM})$ the standard deviation of the dynamic load due to non-suspended masses,
- $\sigma(\Delta R_{SM})$ the standard deviation of the dynamic load due to suspended masses (for details the interested reader should read [140])

and:

$$\bar{\text{A}}_{dyn} = \frac{1}{2\sqrt{2}} \cdot \sqrt[4]{\frac{l^3 \cdot h_{TR}}{E \cdot J}}$$

where h_{TR} the total dynamic stiffness of the track given by:

$$h_{TR} = \frac{1}{2\sqrt{2}} \cdot \sqrt[4]{E \cdot J \cdot \frac{k_{total}}{l}}$$

Where

- E, J the modulus of elasticity and the moment of inertia of the rail
- k_{total} the "rail support modulus" or "total track stiffness (static)" in kN/mm (see par. 16.1),

In the motion of the Non Suspended Masses (NSM) of the vehicle a section of track is also participating. For an accurate calculation of this track mass m_{TRACK} - participating in the motion of the m_{NSM}- a detailed theoretical analysis compared to data from measurements is cited in [133].

In all the theoretical methods above the total static stiffness of track plays a key role: the more elastic the track is, the less the ties are stressed. It is therefore evident that resilient fastenings play a key role in the distribution of loads on track, and eventually in the life-cycle of the track [134].

In conclusion: **Maintaining a good track condition, the wheel / rail contact forces are kept at low level, thus reducing the track wear.**

28 Assessment of Track Geometry Quality

28.1 Overview

The condition of the superstructure is described by

1. The track quality
2. The amount and the severity of speed restrictions (speed decrease over a length)
3. Track positions with long lasting problems

Track geometry is normally measured by track recording cars. They enable the standard deviations to be calculated, which have proved to be useful for describing the track quality. In some cases, also vehicle reactions calculated from the recorded geometry are used to assess track quality.

28.2 Standard Deviation

The standard deviation σ (or SD) is one of the most commonly used track quality indicators (TQIs) by European Railway Networks. It represents the dispersion of the signal (position of the measured point along the track) over a given track section, in relation to the mean signal (mean position of the track) over the considered section.

$$\sigma = \sqrt{\frac{\sum_{i=1}^{N} (x_i - \bar{x})^2}{N-1}}$$

Where

N is the number of samples

$\bar{x}$ is the mean of the signal

σ is the standard deviation

The standard deviation is commonly calculated for the following parameters:

- Longitudinal level D1 (D1: Wavelength range 3 m < $\lambda \leq$ 25 m) (see EN 13848-1)
- Alignment D1

It is also calculated for the following parameters:

- Twist
- Track gauge

K. Tzanakakis: *The Railway Track and Its Long Term Behaviour*, STTT 2, pp. 195–197.
DOI: 10.1007/978-3-642-36051-0_28

- Cross level
- Longitudinal level D2 (D2: Wavelength range 25 m < $\lambda \leq$ 70 m) (see EN 13848-1)
- Alignment D2

Standard deviation for the longitudinal level and alignment is calculated either separately for each rail or combined (for example: mean, worst, best or outer rail, depending on the use to be made of the SD).

Commonly, standard deviation is calculated over a 100 or 200-m length. It may be calculated either at fixed distances without overlap or, with overlap, as a sliding standard deviation. Calculation of standard deviation is also done over longer distances such as 1 km, an entire line or an entire network.

28.3 Isolated Defects

Isolated defects are mainly related to safety; however, counting the number of isolated defects exceeding a specified threshold such as Intervention Limit (IL) or Alert Limit (AL) (see par. 32.9) on a given fixed length of track can be representative of the track geometric quality. This method is used by several European networks.

The number of isolated defects per unit of track length is commonly counted for the following parameters:

- Longitudinal level D1
- Alignment D1
- Twist
- Track gauge
- Cross level

It is also counted for the following parameters:

- Longitudinal level D2
- Alignment D2

Commonly, the number of isolated defects is counted over 1 km or more. It may also be counted over 100 or 200 m of track.

If required distinction between specific track sections can be made, such as plain lines, stations and switches and crossings.

Alternatively a calculation can be made to specify what percentage of a line exceeds a certain threshold level.

28.4 Combination of Various Parameters (Combining Standard Deviations)

Assessment of the overall track geometry quality of a track section (200m, 1000m, ...) can be made by a combination of standard deviations (SDs) of the individual geometric parameters. An example of such a TQI is given below.

$$TQ_i = \sqrt{w_{Al} \cdot \sigma_{Al}^2 + w_G \cdot \sigma_G^2 + w_{Cl} \cdot \sigma_{Cl}^2 + w_{Tp} \cdot \sigma_{Tp}^2}$$

Where:

Al: mean of the alignment of both rails

G: track gauge

Cl: cross level

Tp: mean of the longitudinal level of both rails

w: weighting factor of the geometry parameters

σ: standard deviation of the individual geometry parameters

In this example, the TQI describes an overall track geometric quality and results from a root-mean-square sum of the SDs of individual geometry parameters multiplied by their weighting factors.

It is up to the Infrastructure Manager to determine the weighting factors and TQI limits, e.g. for tamping purposes the weighting factor w_G is zero.

Another method might be to transform the standard deviations of geometry parameters or their combinations into a dimensionless number that can be used without distinction of line category, speed range and track geometry parameter.

Section VI: Design – Construction – Maintenance Aiming at High Performance

Knowledge is power
Francis Bacon

A high performance railway track has to be designed and constructed properly. And then, when the trains are running on the track (operational phase), the track has to be monitored and maintained. What are the possible maintenance strategies? Which strategy best fits the needs and the culture of the various Infrastructure Managers?

29 Track Design

Key terms:

Allowable stress design, service life of track, track condition monitoring, monitoring optimisation, track maintenance, asset management, maintenance strategies, effective (rational) maintenance, maintenance policies, track maintenance planning, track segmentation, maintenance actions

29.1 Some History of Design of Structural Elements

For many centuries, buildings and other structures were designed using common sense, trial and error, and rules of proportion acquired through experience. Their effectiveness depended on the knowledge and skills of master craftsmen.

In the nineteenth century industrialization and the mass-production of iron and steel led to rapid changes in construction methods. This in turn provided a motivation to replace the traditional trial-and-error approach for designing structures, by calculations based on scientific principles. The only scientific tools available at that time for designing structures were Newton's laws of motion and the theory of elasticity. As time went on, these scientific principles were developed into a unified, practical tool for structural calculations called allowable stress design [9].

Time period	Designing methods
Before 19th century	Common sense, trial and error, and rules of proportion acquired through experience
19th century (Industrialization and the mass-production of iron and steel)	Calculations based on scientific principles: • Newton's laws of motion and • theory of elasticity in time developed to
Now	Allowable stress design

29.2 Allowable Stress Design

In allowable stress design, the capability of a structural element is checked by calculating the elastic stresses in it due to the maximum expected loads, and

K. Tzanakakis: *The Railway Track and Its Long Term Behaviour*, STTT 2, pp. 201–203.
DOI: 10.1007/978-3-642-36051-0_29

comparing them with allowable stresses. The allowable stress is equal to the failure stress of the material divided by a safety factor. Safety factors were first determined by applying allowable stress design methods to successful structures existing at that time. The safety factors for new materials were estimated in comparison with those for traditional materials by taking into account the nature of failure for the new material and its uncertainty or variability. Allowable stress design formed the basis of structural codes and standards, including track design standards, for most of the twentieth century.

For the railway structural elements is the track itself. An allowable stress design flow chart for conventional ballasted track structure design is typically used by the track design engineer to undertake design of new railway track. This flow chart is shown in Figure 123 [9].

The traditional method for designing new railway track uses an "allowable stress" design approach in determining the load applied to the track. This approach expresses the wheel load empirically as a function of the static wheel load with a dynamic factor to account for vehicle and track irregularities.

$$P_D = \varphi \cdot P_0$$

where:

P_D = design wheel load (kN);

P_0 = static wheel load (kN); and

φ = design dynamic increment (always > 1).

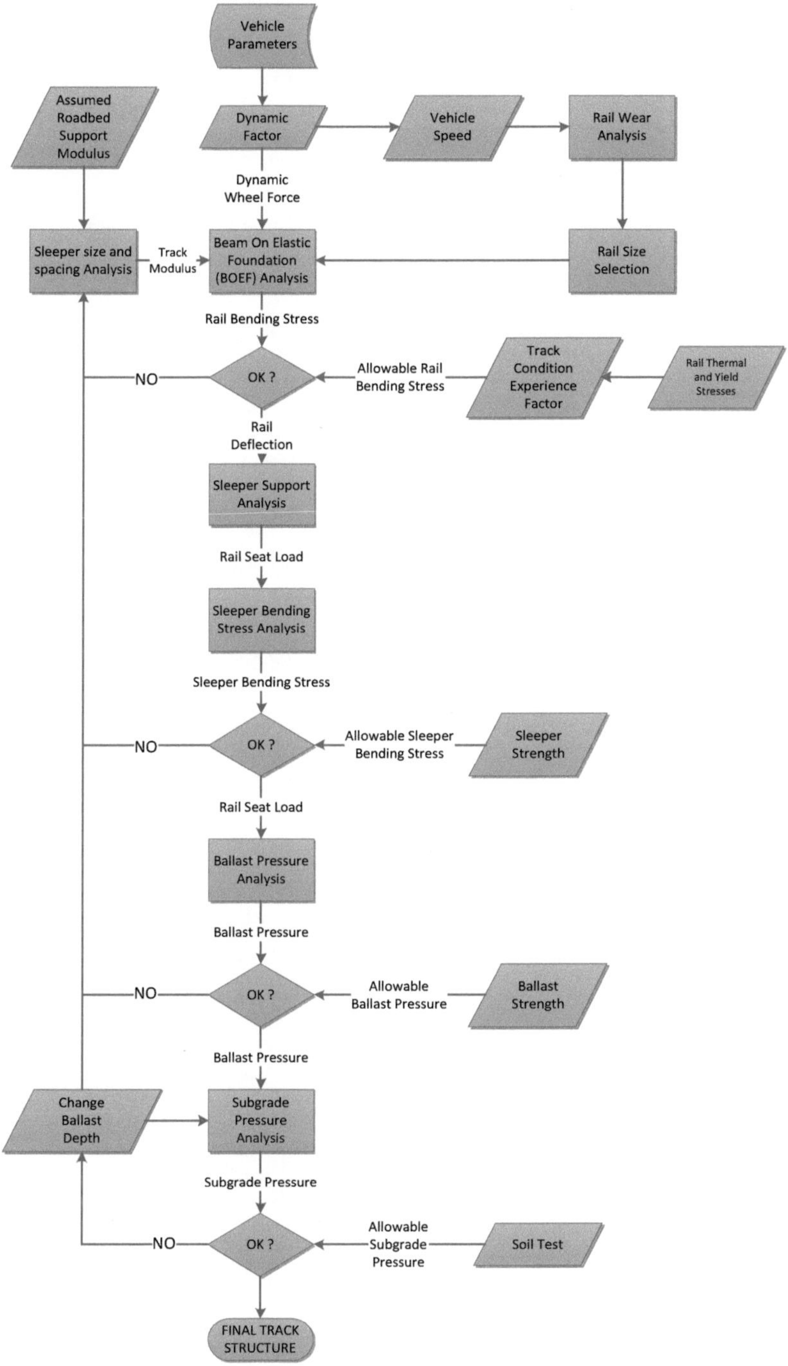

Fig. 123 Track Design Procedure (after Prause et al 1974) [9]

30 The Track Construction (Aiming at High Performance)

Fig. 124 Early railway track construction

30.1 The Service Life of the Track

As we examined in Chapter 24, track quality is deteriorating due to traffic, showing various faults. The development of the track quality is shown in Figure 113.

Let's examine the track behaviour over time:

In time t=0 the track has the initial quality Q_0 (usually its best quality) (Figure 125). The track begins to deteriorate, so on time t_1, at reaching the intervention limit[47], maintenance actions are undertaking, improving the track quality at a level Q_1 ($Q_1 < Q_0$). At time t_2, the intervention limit is reached again. New maintenance actions improve the track quality at a level Q_2 ($Q_2 < Q_1 < Q_0$). These maintenance cycles repeats at smaller and smaller intervals (t_n-t_{n-1} <.... < t_1-t_0). When the intervals become very short, the track must be replaced, because the maintenance cost are very high for maintaining the initial performance level (or the performance level must be decreased): the service life of the track is at its end.

The description of the service life of the track is presented very simplified. In reality, every track component has its own (usual) service life (par. 30.2).

[47] Alert limit or immediate action limit, according the respective maintenance strategy, see par. 32.9.

K. Tzanakakis: *The Railway Track and Its Long Term Behaviour*, STTT 2, pp. 205–207.
DOI: 10.1007/978-3-642-36051-0_30

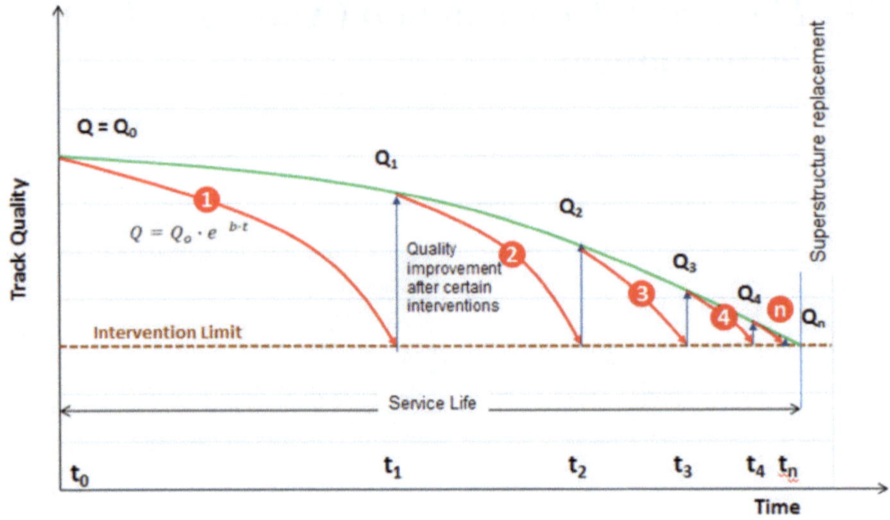

Fig. 125 The service life of a track

30.2 Typical Life Cycles

On a typical intensively operated main line the following typical service life apply:

Table 22 Typical life cycles [11]

Action on the track	Traffic load	Frequency
Tamping	40-70 mgt	4-5 years
Grinding	20-30 mgt	1-3 years
Ballast cleaning	150-300 mgt	12-15 years
Rail renewal	300-1000 mgt	10-15 years
Timber sleeper renewal	250-600 mgt	20-30 years
Concrete sleeper renewal	350-700 mgt	30-40 years
Fastenings renewal	100-500 mgt	10-30 years
Ballast renewal	200-500 mgt	20-30 years
Formation renewal	> 500 mgt	> 40 years

On track of the highest category, a work cycle of up to 8 years is the norm. Ballast cleaning is undertaken only in connection with other major treatment (on timber sleepered track about every 20 years, on concrete sleepered track about every 30 years). The service life of rails is usually half the service life of the track system about 10 years): the limiting factor is not usually that the rail head is fully worn, but the incidence of metallurgical faults in the running surface. Ballast renewal normally takes place in connection with full track relaying (at between 20 and 30 years intervals) [11].

31 Track Condition Monitoring

31.1 Introduction

The Infrastructure Manager aims at a high quality track not only when the track is new (high initial quality), but also during its life. In order to take the right decisions for maintenance he must have

- right standards for monitoring the track condition
- proper track condition analysis tools
- right track condition data's (information) to be analysed, in order to understand the track behaviour.
- processes to assure the proper working of the whole monitoring system.

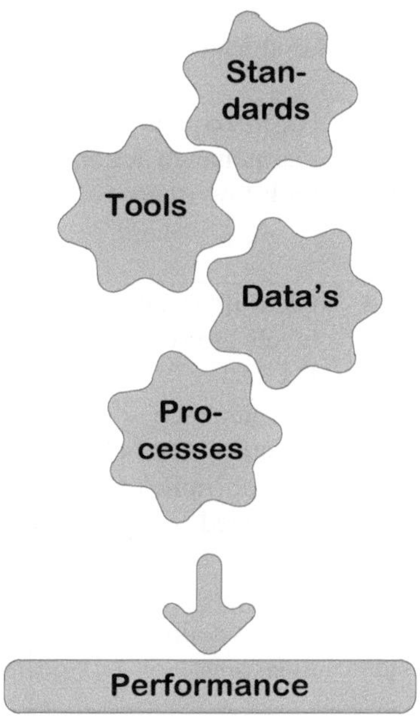

Fig. 126 Track condition monitoring for high performance

Through a proper analysis of the track condition and behaviour using historical data's (of an adequate quality level), a failure prediction, and also predictive maintenance as well as condition- based maintenance can be achieved and optimised (see Chapter 32).

Condition monitoring data is expected to increase in importance in the next decade, which will make necessary a higher quality of data collection, analysis and use.

K. Tzanakakis: *The Railway Track and Its Long Term Behaviour*, STTT 2, pp. 209–215.
DOI: 10.1007/978-3-642-36051-0_31

31.2 Definition of Infrastructure Monitoring

The UIC Asset Management Working Group [97] gives the following definition of infrastructure monitoring:

> *"Infrastructure Monitoring is the continuous or periodical measurement of the condition of infrastructure components and/or subsystems".*

Monitoring systems can provide information on the condition or use of the infrastructure.

31.3 The Monitoring as Part of the Maintenance Strategy

Monitoring is part of the maintenance strategy of the Infrastructure Manager (IM), and varies depending on the needs and the respective maintenance strategy, as presenting in the following examples [97]:

Maintenance Strategy	Maintenance actions	Monitoring System (track condition information)
Example 1: Very small deteriorations of the track geometry allowed.	• High-frequency maintenance actions with low-impact	• Monitoring at high frequency • High accuracy of the information
Example 2: Maintenance strategy based on a good knowledge of the track geometry deterioration	• Clear intervention levels are set by the IM • Maintenance executed just before danger limits are exceeded	• Monitoring only at intervals in the order of magnitude of deterioration time, • the required accuracy can be slightly lower than in Example 1.
Example 3: No clear maintenance strategy and / or no deterioration analysis is possible or available	• timely maintenance measures	• high accuracy and condition information in small time intervals may be needed

Maintenance strategies will be further discussed in Chapter 32.

31.4 The PPIMS, a Top-Down Method to Optimise Monitoring

The UIC Asset Management Working Group [97] proposed a top-down approach for track condition monitoring (Figure 127). This approach was named PPIMS: The letters stand for Performance / Process / Information / Monitoring / Standard.

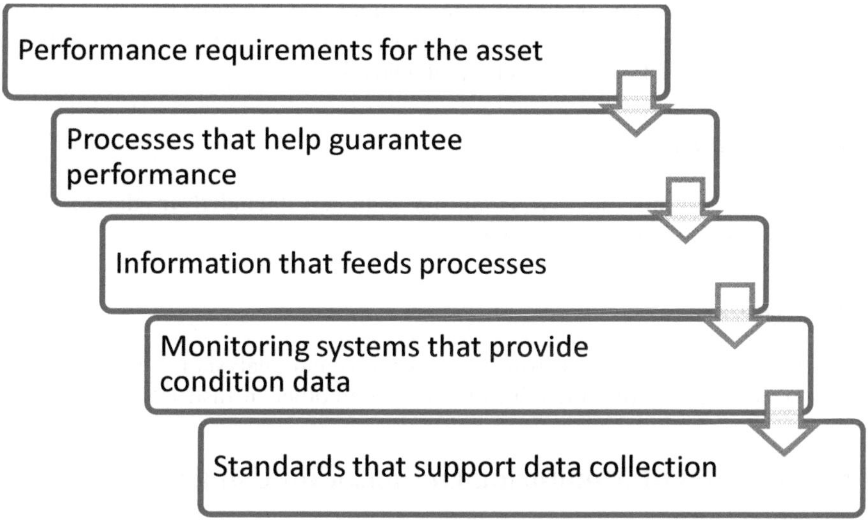

Fig. 127 A top-down approach for track condition monitoring

This method supports the idea that, in order for Infrastructure Managers to improve their performance, they must take the right decisions and control their processes, supported by the right information analysed intelligently, based on the right monitoring data collected wisely, and applying the right standards [97].

31.4.1 Performance

It's very important for the Infrastructure Managers to know the performance requirements of the track and its components. It's also important to know the failure modes, impact of failures and to understand the track deterioration and finally performance killers and cost drivers.

It is essential to recognize how each critical component or system contributes to overall company performance.

31.4.2 Process

Processes are crucial to manage the performance of the track and its components and help Infrastructure Managers meet their track performance targets.

Examples of processes are: restoring the track geometry, maintaining switches, grinding the rails, renewing welds, and cleaning ballast.

31.4.3 Information

For the processes the optimum information is needed.

Optimum information include: precise condition parameters, the desired accuracy, the frequency of measurement (number of times per year), the accuracy of the measurement data and the method of analysis.

31.4.4 Monitoring

Methods of collecting the raw condition data are included in the technical guidelines of the Infrastructure Manager.

31.4.5 Standards

A further standardisation of measurement methods and analysis is needed in order to apply benchmarking (compare information) with other Infrastructure Managers.

31.5 Examples of Monitoring Optimisation Using PPIMS

31.5.1 Example 1: Rolling Contact Fatigue

Rolling contact fatigue (RCF) phenomenon is analytically examined in par. 19.2 and 19.3.2.

In the study of the UIC Asset Management Working Group [97], a PPIMS analysis of the phenomenon and the use of appropriate counter-measures have been examined, as presented below.

31.5.1.1 Performance

The negative impacts of RCF on the performance are connected to:

- safety (derailment risk),
- reliability (signalling failure upon rail break) and
- cost (expensive repair works).

31.5.1.2 Process, Information and Monitoring

The countermeasures (processes) are preventive and corrective rail grinding, rail replacement and rail lubrication.

Process 1	**Grinding (preventive and corrective)**
Information	The frequency and the grinding sections are often determined by estimation and experience. Ideally these expert knowledge estimates should be based on extensive deterioration and track use information. Once deterioration patterns are known, the early warning indicators can be chosen and monitored. Track tonnage information is becoming increasingly available through axle load checkpoint measurements. Deterioration information requires more attention.
Monitoring	Head checks: visual inspection and eddy current measurement techniques are used. Squats and other RCF cracks: conventional Ultrasonic Inspection Vehicle (UIV) or portable equipment is used.

Process 2	**Rail replacement (preventive and corrective)**
Information	Preventive rail replacement: expert knowledge and track section usage information are required. Corrective rail replacement: reliable information about the actual condition of the track must be collected through monitoring.
Monitoring	The minimum rail cross-sectional profiles has to be measured at intervals determined in the technical guidelines of the Infrastructure Manager.

31.5.1.3 Standards for RCF

There are not any international standards.

31.5.2 Example 2: Switches and Crossings (S&C)

The following failures are the main ones affecting the performance of S&C [97]:

- Excessive force applied to move the switch blade,
- Closing signal to the interlocking system not within the tolerance range,
- Bad alignment and levelling,
- Cracks initiated on the surface or sub-surface (RCF), or internal cracks,
- Corrugation and wear,
- Ballast deterioration, voids,
- Joint deterioration,
- Loose or missing fasteners,
- Shift in neutral temperature (stress build-up in rail).

These failures lead to a risk of derailment, high maintenance costs, and low reliability and availability.

Countermeasures (processes) used by Infrastructure Managers are:

- Lubrication of the switch blade,
- Adjustment of switch settings,
- Restoration of the rail surface by welding,
- Tamping or ballast cleaning,
- Grinding,
- Replacement,
- Neutralization of the stress build-up.

31.5.3 Example 3: Track Geometry

Track geometry is an important performance parameter for the safe availability of the network. Like all safety- related parameters, geometry faults are normally kept at a level ensuring that the risk of derailment is kept low (see par. 26.4.2).

Keeping the risk so low makes track geometry maintenance a cost driver, so track geometry recording is a main monitoring issue.

Processes for restoring the track geometry are: tamping, ballast cleaning and renewal, sleeper replacement, joint repair, rail neutralisation, rail replacement, rail grinding and substructure treatment.

31.6 Parameters Usually Measured for Track Condition Monitoring

For monitoring the track condition, the following parameters are usually measured:

31.6.1 Track Measurement

Track Geometry	• Left/right longitudinal levels • Left/right alignments • Curvature • Gauge • Cant • Twist (2m, 9m or user selected base)
Rail Profile	• Vertical wear • Horizontal wear • Wear 45°
Rail Corrugation	• Rail corrugation in several wavebands in the range between 20 and 3000 mm

Surface Defects	• Rail breakage • Abrasion from skidding • Shelling • Flaw • Corrugation • Welding anomalies • Rubble imprints • Absence of fastener • Cracks in the sleepers Integrity of sleepers fasteners

31.6.2 Ride Quality

Wheel-Rail Interaction Forces Measuring System	• Real time Y & Q force monitoring • Real time Y/Q ratio monitoring • Bogie lateral acceleration correlation
Vehicle Accelerations	• Comfort and RMS[48] value index calculation, statistical analysis according to UIC Leaflet 518 [70]
Wheel-Rail Contact Geometry	• Angles of contact and equivalent conicity at several values of standard deviation σ.

[48] Root mean square, is a statistical measure of the magnitude of a varying quantity.

32 Track Maintenance

32.1 Introduction

Investments on the railway track refer to assets with very long life. Today many tracks are over 100 years old. Of course some track elements are replaced during maintenance actions over the years, but other track elements might remain the same – especially the substructure. In any case, a large amount of maintenance is necessary to ensure this long life.

Maintenance has the following main targets for the respective issues, as presented in the following table [47]:

Table 23 The targets of maintenance

Issue	Description	Target
Safety	Probability for accidents	• As low as possible.
Comfort	Comfort refers both to passengers and freight as well as to the environment (noise and vibration).	• Highest comfort for passengers and freight, • minimum impacts to the environment
Availability	The availability of the track depends on failures and speed restrictions.	• Availability as much as possible

K. Tzanakakis: *The Railway Track and Its Long Term Behaviour*, STTT 2, pp. 217–249.
DOI: 10.1007/978-3-642-36051-0_32

Table 23 (*continued*)

Economy	A low quality track is a very costly track, since the track deterioration affects both track and trains. At the same time maintenance is expensive.	• A cost effective track

Optimization of maintenance is needed to have a track of high performance. For optimal maintenance the task of highest priority and importance is to **make the right measure at the right time** to fulfil the requirements of safety, comfort and availability in terms of LCC (see SECTION VII). This task requires complete knowledge about the current condition of the track and also the complete knowledge about the influence and the effect of the different types of maintenance on the track.

The goal of effective modern maintenance strategies is to come as close as possible to this optimum. Decisions about maintenance are based today on

- measurements of important parameters which are analysed to give knowledge about the condition of the track
- knowledge about the deterioration behaviour of the track (from models and/or skilled engineers)
- regulations,
- budget constraints.

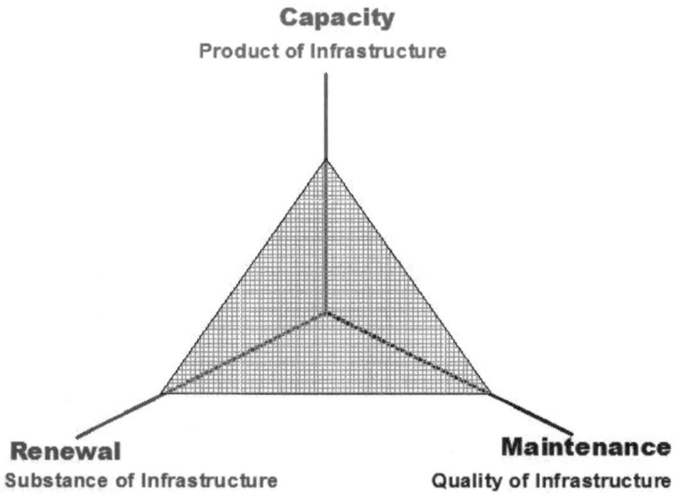

Fig. 128 Variables of railway infrastructure management (see also 25.1)

Optimal maintenance can add life to the track, saving lots of money. Taking into account budgetary constraints and since such a large part of budget of the Infrastructure Manager is spent on maintenance, it is of vital importance to have a clear and effective maintenance strategy.

The main objective of successful policy is to define maintenance and renewal rules leading to a sustainable balance between the magnitude of the three main decisive variables of infrastructure properties (Figure 128) (see also Chapter 25):

- Maintenance (Track quality is mainly determined by maintenance)
- Capacity (Capacity is a "product" of infrastructure)
- Renewal (the substance of a railway network relates directly to its renewal cycle)

32.2 A Historical Background for Maintenance

32.2.1 The Transition of a Manual-Labour-Based Economy to Machine-Based Manufacturing

In the early times when wooden and stone tools were made, the execution of certain works aimed to keep them in a good working condition. Maintaining those tools required nothing more than very elementary skill.

Fig. 129 Axe heads found at a 2700 BC Neolithic manufacture site in Switzerland, arranged in the various stages of production from left to right.

In the later part of the 1700s the previously manual-labour-based economy transformed gradually into machine-based manufacturing. It started in Great Britain with the mechanization of the textile industries, the development of iron-making techniques and the increased use of refined coal [101].

Trade expansion was enabled by improved roads and railways. Steam power and powered machinery (mainly in textile manufacturing) underpinned the dramatic increases in production capacity.

Behind the industrial revolution the driving force was specifically the double-acting Watt steam engine, developed between 1763 and 1775 (Figure 130).

Fig. 130 Watt's steam engine at the lobby of the Higher Technical School of Industrial Engineering of Madrid

32.2.2 *The Industrial Revolution and the Need for Inspection and Maintenance*

The development of high-pressure steam engines started with Richard Trevithick[49], from around 1800. High steam pressures brought about the need for increased safety awareness, resulting in periodic inspections to assess general condition and integrity.

Fig. 131 Preserved British steam-powered fire engine – an example of a mobile steam engine. This is a horse-drawn vehicle: the steam engine drives the water pump.

[49] Richard Trevithick (13 April 1771 – 22 April 1833) was a British inventor and mining engineer from Cornwall. His most significant success was the high pressure steam engine and he also built the first full-scale working railway steam locomotive. On 21 February 1804 the world's first locomotive-hauled railway journey took place as Trevithick's unnamed steam locomotive hauled a train along the tramway of the Penydarren Ironworks, in Merthyr Tydfil in Wales.

Maintenance became more important during the Industrial Revolution. The equipment manufactured during the industrial revolution was simple in comparison to today's equipment, and intended to **last for a lifetime**.

Patterns of wear and deterioration were simple, and rates of wear were predictable. It was indeed possible for a skilled worker to understand these equipment so well that he knew exactly how long specific components would last. Of course unexpected failures occurred, as manufacturing techniques, quality standards and especially metallurgy were not well developed. Few of these machines were built, and it was common to build these according to the customers' needs – standardization in terms of parts was not the rule.

Maintenance of this equipment relied greatly on the skill of the worker. He had to rebuild parts, and often had to manufacture parts from only a sample. It is for these reasons that the old blacksmith was a popular and prominent person.

The Industrial Revolution brought some very sophisticated equipment about, for example printing presses and the Jacquard weaving mill. These machines were examples of mechanical geniality, and were manufactured with great skill and precision. Maintaining these machines required exceptional skill from the workers.

Fig. 132 Jacquard Looms in Masson Mill

Maintenance Management, or Physical Asset Management involved only the "people management" and a certain degree of quality control.

Successful maintenance was based on **skill**.

32.2.3 Starting the Mass Production

In the first half of the twentieth century the mass production started, with more strong competition. This economy transformation had the following parameters:

- need to increase profit led to the need to decrease operating cost (larger volumes had to be produced at lower cost),
- need to extend the life of production equipment
- pressure to increase the production.

- equipment became larger, faster, more automated and more complex.
- plants were required to produce more reliable (fewer interruptions) and for longer periods
- capital for investments was required to set up or expand a plant

During this period, needs beyond worker skill developed. To execute maintenance tasks quicker, more effective and correctly, a high level of technical knowledge was required.-

The dominating maintenance policy in the time period up until the Second World War, was run to failure [101] (see par. 32.6.1).

Successful maintenance was based on **skill and technical knowledge**.

32.2.4 Period 1945-1960

During the time period up until the 1960s safety matters became more important as well as improvement in labour efficiency, and a more preventive maintenance strategy emerged.

The competition of the Americans and Russians led inter alias to the space race, with the success of the various space programs heavily dependent upon **reliable operation** of equipment for given periods.

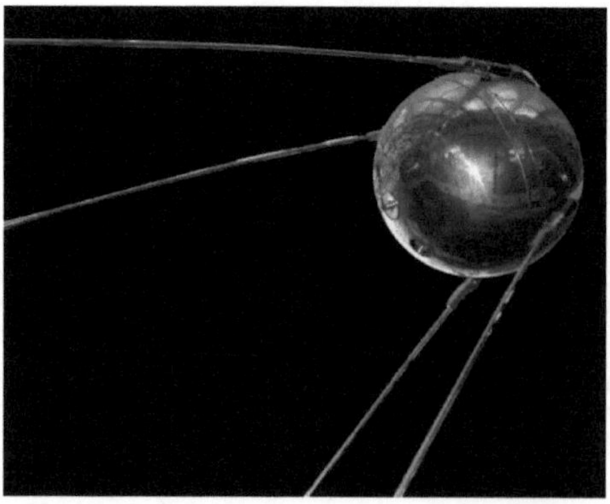

Fig. 133 The launch of Sputnik 1 marked the start of the Space Age (on October 4, 1957)

Fig. 134 Launch of Apollo 11 spacecraft to the moon on July 16, 1969

Failure prevention became critical to ensure continued output and reduce cost and losses. Economic viability of a plant became dependent on the capability to predict the failure, and to prevent it by doing the relevant maintenance during off-peak periods.

The relative cost of maintenance became significant, and attempts were made to contain it by better management skills.

In the period 1945 -1960, maintenance was characterized by a need for skill, technical knowledge, and **preventive and planning capabilities**.

32.2.5 The Period 1960 - Now

Since the 1960's, the concept of "Reliability Centered Maintenance" (RCM) was initiated by the "Federal Aviation Agency" in the USA.

There was a need for a "total approach" to maintenance – in other words the examination of what is

- the most appropriate mixture of maintenance types
- for a given equipment type,
- in a given operating context.

This "most appropriate mixture of maintenance types" is often called the "maintenance strategy", **applying the most appropriate mixture of preventive, detective, corrective and adaptive tasks in order to get to the most favourable balance between cost and availability / reliability**.

32.2.6 The Need to a Physical Asset Management

A number of disasters forced industry and regulatory authorities to seriously rethink what the minimum requirements for good physical asset management are. The Piper Alpha disaster in the North Sea is perhaps the most notable (Figure 135). On July 6, 1988, as part of routine maintenance, technicians checked the good condition of the valves. Unfortunately, the technician made a mistake and forgot to replace one of them. At 10 pm on the same day, a technician began pressing the button to move the liquid pumps and gas most expensive in the world, and simultaneously started a horrible accident.

In less than two hours, the fire threw down the entire contents of the building killed 167 workers and caused the loss of $ 3.4 Billion. Only 62 crew members survived out of the 229 personnel on board. Many workers were trapped because rescue helicopters could not approach when confronted with flames 100 meters high. The majority of those who survived jumped from the platform into the rough sea.

A public inquiry conducted by the British government under Lord Cullen and performed combustion/fire tests to investigate possible sources of ignition. Lord Cullen's recommendations were either taken up in legislation or became industry standards.

Fig. 135 The Piper Alpha disaster

So **Publicly Available Specification number 55 (PAS 55) was born**, with the first release published during April 2004.

PAS 55 is an asset-centric approach to physical asset management, with risk management very much a central theme, and it pulls all the "levels" of physical asset management together into one integrated stratified system.

32.4 Track Maintenance and Asset Management

32.4.1 What Is Asset Management?

Asset management is defined in [99] as:

> *"systematic and coordinated activities and practices through which an organization optimally and sustainably manages its assets and asset systems, their associated performance, risks and expenditures over their life cycles for the purpose of achieving its organizational strategic plan*

Organizational strategic plan is defined in [99] as:

> *"overall long-term plan for the organization that is derived from, and embodies, its vision, mission, values, business policies, stakeholder requirements, objectives and the management of its risks"*

The PAS 55 specification can provide guidelines and recommendations on "WHAT TO DO" although it does not prescribe solutions on how to do it.

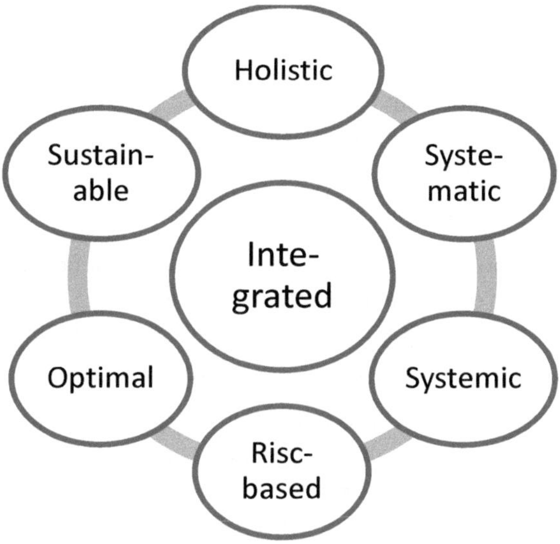

Fig. 136 Key principles and attributes of asset management [99]

The key principles and attributes of asset management are shown in Figure 136 and explained in the following table [99]:

holistic	looking at the whole picture, rather than a compartmentalized approach
systematic	a methodical approach, promoting consistent, repeatable and auditable decisions and actions
systemic	considering the assets in their asset system context and optimizing the asset systems value (including sustainable performance, cost and risks) rather than optimizing individual assets in isolation
risk-based	focusing resources and expenditure, and setting priorities, appropriate to the identified risks and the associated cost/benefits
optimal	establishing the best value compromise between competing factors, such as performance, cost and risk, associated with the assets over their life cycles
sustainable	considering the long-term consequences of short-term activities to ensure that adequate provision is made for future requirements and obligations (such as economic or environmental sustainability, system performance, societal responsibility and other long-term objectives)
Integrated	recognizing that interdependencies and combined effects are vital to success. This requires a combination of the above attributes, coordinated to deliver a joined-up approach and net value.

Effective implementation of asset management requires a disciplined approach which enables an organization to

- **maximise value** and
- **deliver its strategic objectives** through managing its assets over their **whole life cycles**.

This includes

- determination of appropriate assets to acquire or create in the first place,
- how best to operate and maintain them, and
- adoption of optimal renewal, decommissioning and/or disposal options.

The principal benefits of optimized life cycle asset management include:

- enhanced customer satisfaction from improved performance and control of product or service delivery to the required standards;
- improved health, safety and environmental performance;
- optimized return on investment and/or growth;
- long-term planning, confidence and performance sustainability;
- the ability to demonstrate best value-for-money within a constrained funding regime;
- evidence, in the form of controlled and systematic processes, to demonstrate legal, regulatory and statutory compliance;
- improved risk management and corporate governance and a clear audit trail for the appropriateness of decisions taken and their associated risks;
- improved corporate reputation, the benefits of which may include enhanced shareholder value, improved marketability of product/service, greater staff satisfaction and more efficient and effective procurement from the supply chain;
- the ability to demonstrate that sustainable development is actively considered within the management of the assets over their life cycles.

Such benefits are most often associated with the application of a number of key principles and attributes in the design and implementation of the asset management system (Figure 136)[50].

The following enabling elements are also considered to be essential for the successful implementation of these principles (Figure 137):

- an **organizational structure** that facilitates the implementation of these principles with clear direction and leadership;
- **staff** awareness, competency, commitment and **cross-functional coordination**;
- adequate **information** and knowledge of **asset condition, performance, risks and costs**, and the interrelationships between these.

[50] These are explained further in PAS 55-2.

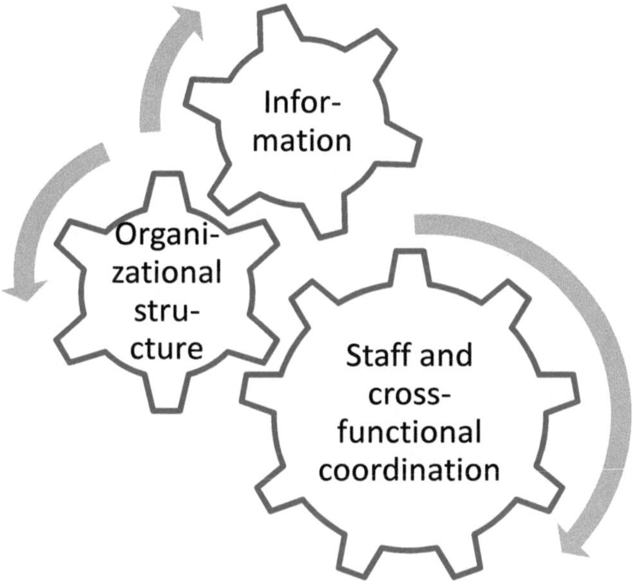

Fig. 137 Elements of successful implementation of asset management

32.4.2 The Structure of an Asset Management System

An asset management system is primarily designed to support the delivery of an organizational strategic plan (see par. 32.4.1) in order to meet the expectations of a variety of stakeholders[51] (Figure 138) [99].

The organizational strategic plan is the starting point for development of the

- asset management policy,
- strategy,
- objectives and
- plans.

These, in turn, direct the optimal combination of life cycle activities to be applied across the diverse portfolio of asset systems and assets (in accordance with their criticalities, condition, performance and chosen risk profile of the organization).

32.4.3 The Asset Management and the Infrastructure Manager

The PAS 55 specification is rapidly gaining ground as the standard of choice for the management of physical assets. It does not define "the ideal" asset management system, but rather provides a set of minimum requirements to which the asset management system should conform.

[51] Stakeholders, see Chapter 8.

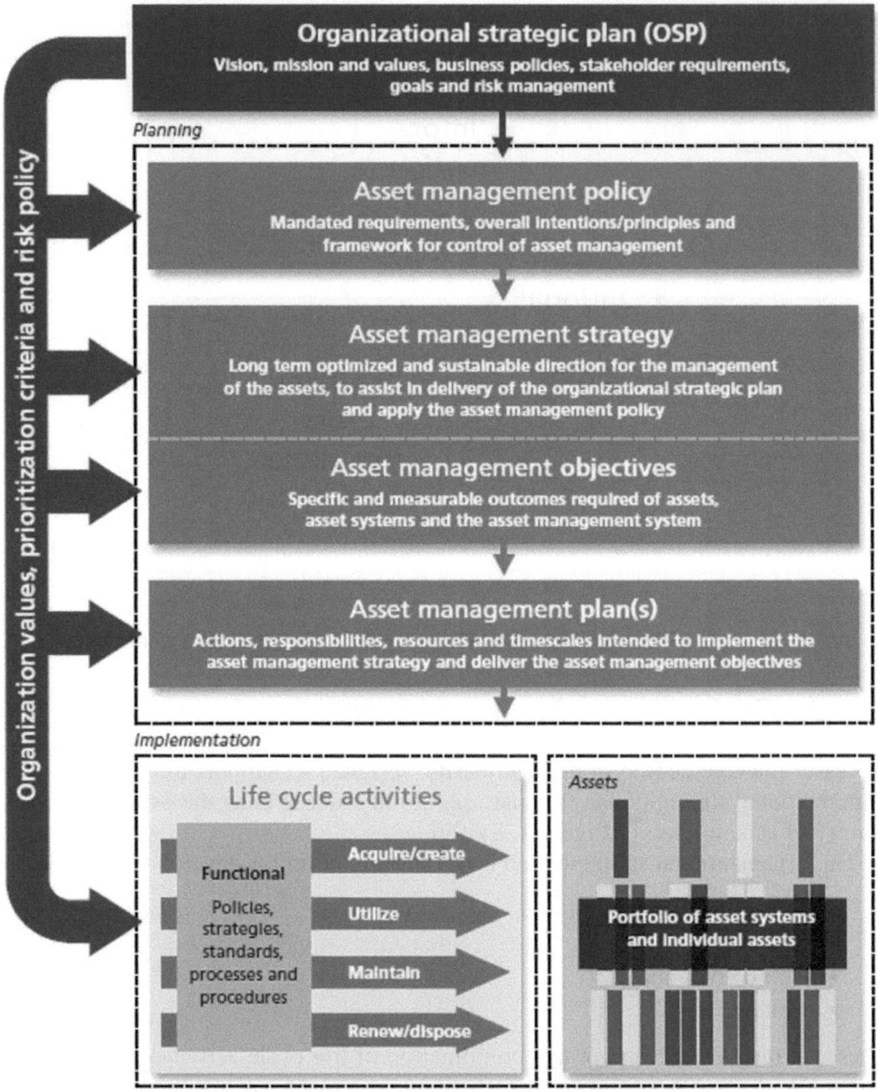

Fig. 138 Planning and implementation elements of an asset management system [99]

The Infrastructure Manager has to balance between cost and performance (Figure 139).

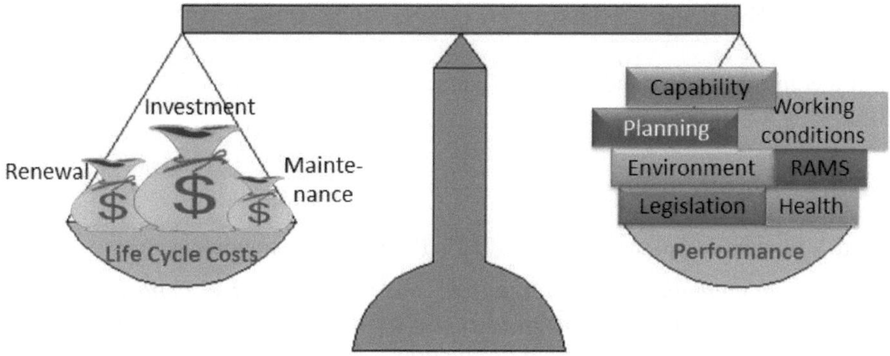

Fig. 139 Managing the balance (UIC Asset Management Working Group)

> If you don't apply asset management you will end up being considered either too expensive or not good enough[52].

32.4.4 Integration with Other Management Systems

The implementing of the PAS 55 specification would enable an organization to align or integrate its asset management system with other related management systems. In order to enable organizations to develop integrated management systems, the requirements and structure of this PAS are arranged within the Plan-Do-Check-Act (PDCA) framework as follows (see Figure 140):

Plan	establish the asset management strategy, objectives and plans necessary to deliver results in accordance with the organization's asset management policy and the organizational strategic plan
Do	establish the enablers for implementing asset management (e.g. asset information management system(s)) and other necessary requirements (e.g. legal requirements) and implement the asset management plan(s)
Check	monitor and measure results against asset management policy, strategy objectives, legal and other requirements; record and report the results
Act	take actions to ensure that the asset management objectives are achieved and to continually improve the asset management system and asset management performance

Fig. 140 The Plan - Do - Check - Act framework

52 UIC, Infrastructure Asset Management Seminar, Paris, 19th May 2009 (Asset Management Working Group).

32.5 Definitions

For a common understanding it is of great importance to provide definitions of the following key terms.

32.5.1 Maintenance

British Standard Glossary of terms (3811:1993) defines maintenance as:

> *"the combination of all technical and administrative actions, including supervision actions, intended to retain an item in, or restore it to, a state in which it can perform a required function".*

Another definition for maintenance is the following

> *"maintenance is a set of organised activities that are carried out in order to keep an item in its best operational condition with minimum cost acquired".*

32.5.2 Maintenance Strategy

"Maintenance strategy" can be described as *"an arrangement of the "types" of maintenance, which is to be done to the track and its components, at a specific rate or frequency, under specific operating conditions".*

32.5.3 Maintenance Plan

The maintenance plan presents *"a description of the preventive and predictive maintenance and inspection tasks to be performed at maintenance objects. The maintenance plans describe the dates and scope of the tasks".*

A maintenance plan in the rail infrastructure sector is a yearly schedule defining when, where and by whom (Infrastructure Managers and/or contractor) a specific type of rail infrastructure maintenance has to be carried out.

The maintenance planning is only related to renewal activities and preventive / predictive maintenance, since corrective maintenance cannot be planned in advance [20].

32.5.4 Maintenance Planning Process

Maintenance planning process is *"the process of identifying maintenance needs, its own and the contractor capacities, material availability and available budget in order to provide useful and realistic data input for rail infrastructure maintenance planning systems".*

32.5.5 Upgrading

In EU Directive 2008/57[53] *"upgrading means any major modification work on a subsystem or part subsystem which improves the overall performance of the subsystem".*

32.5.6 Renewal

In EU Directive 2008/57[54] *"renewal means any major substitution work on a subsystem or part subsystem which does not change the overall performance of the subsystem"*

The term "renewal" includes *"all activities involved in replacing a rail infrastructure part or object by a same or similar type of rail infrastructure part or object".*

32.6 Types of Maintenance Strategies

There are many general types of maintenance strategies, such as (see Figure 141):

- Run to Failure Maintenance (RTF)
- Preventive Maintenance (PM) [55]
- Corrective Maintenance (CM)
- Predictive Maintenance (PDM)

Any combination of these strategies can apply in practice to the asset (the track). The process of "**developing a maintenance strategy**" is all about:

- Defining an intelligent and functional way of determining the appropriate types of maintenance,
- Determining how often maintenance is to be done to our asset.

These strategies would best contribute to the successful realization of the company's goals.

As shown in the above figure "Inspection" is seen as a separate activity[56].

---

[53] DIRECTIVE 2008/57/EC OF THE EUROPEAN PARLIAMENT AND OF THE COUNCIL of 17 June 2008 on the interoperability of the rail system within the Community.

[54] DIRECTIVE 2008/57/EC OF THE EUROPEAN PARLIAMENT AND OF THE COUNCIL of 17 June 2008 on the interoperability of the rail system within the Community.

[55] Also known as time- or interval-based maintenance.

[56] Generally, inspection can be carried out on before, during or after other maintenance activity [EN 13306:2001].

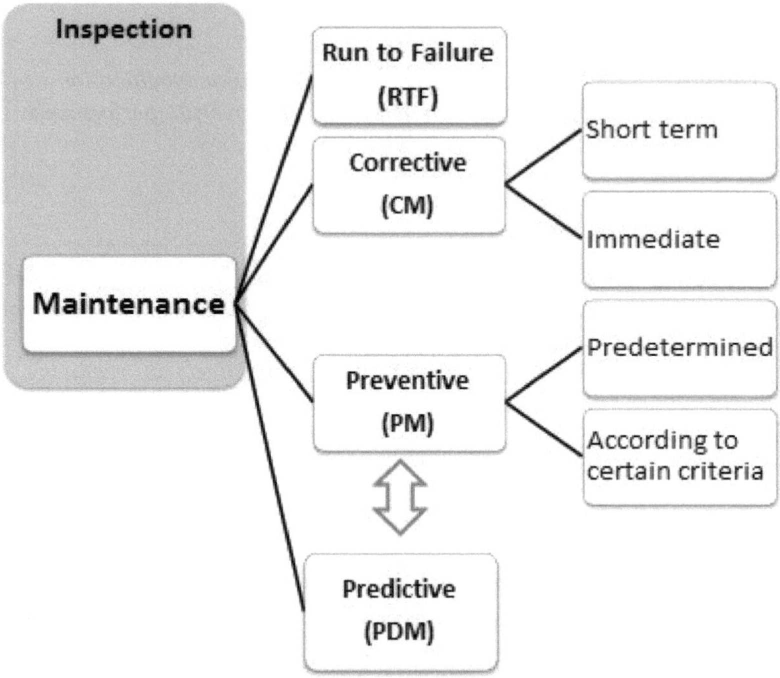

Fig. 141 Breakdown of different maintenance strategy types

32.6.1 Run to Failure Maintenance (RTF)

It is the oldest type of maintenance.

The logic of run to failure management is simple and straightforward: ***When a facility or an asset breaks down, fix it.***

It includes the required repair, replacement, or restore action performed on a machine or a facility **after the occurrence of a failure** in order to bring this machine or facility to at least its minimum acceptable condition.

Disadvantages:

1. Its activities are expensive in terms of both direct and indirect cost.
2. Using this type of maintenance, the occurrence of a failure in a component can cause failures in other components in the same equipment, which leads to low production availability.
3. Its activities are very difficult to plan and schedule in advance.

This type of maintenance is useful in the following situations:

1. The failure of a component in a system is unpredictable.
2. The cost of performing run to failure maintenance activities is lower than performing other activities of other types of maintenance.
3. The equipment failure priority is too low in order to include the activities of preventing it within the planned maintenance budget.

32.6.2 Preventive Maintenance (PM)

Preventive Maintenance is *"a set of activities that are performed on plant equipment, machinery, and systems before the occurrence of a failure in order to protect them and to prevent or eliminate any degradation in their operating conditions"*.

British Standard 3811:1993 Glossary of terms defines preventive maintenance as:

> *"the maintenance carried out at* ***predetermined intervals*** *or according to* ***prescribed criteria*** *and intended to reduce the probability of failure or the degradation of the functioning and the effects limited".*

By this type of maintenance, work is performed out at predefined intervals on a routine basis regardless of whether functionality or performance of the asset is degraded. As the assets age, the frequency and number of checkpoints may need to be re-evaluated. These tasks require a relatively constant amount of labour and materials.

The factors that affect the efficiency of this type of maintenance are:

1. The need for an adequate number of staff in the maintenance department in order to perform this type of maintenance.
2. The right choice of production equipment and machinery that is suitable for the working environment and that can tolerate the workload of this environment.
3. The required staff qualifications and skills, which can be gained through training.
4. The support and commitment from executive management to the PM programme.
5. The proper planning and scheduling of PM programme.
6. The ability to properly apply the PM programme.

32.6.3 Corrective Maintenance (CM)

In the British Standard 3811:1993 Glossary of terms, corrective maintenance is defined as:

> *"the maintenance carried out after recognition and intended to put an item into a state in which it can perform a required function".*

In this type, actions such as repair, replacement, or restore will be carried out after the occurrence of a failure in order to eliminate the source of this failure or reduce the frequency of its occurrence.

Corrective maintenance has several requirements in order to be carried out effectively:

1. Accurate identification of incipient problems.
2. Effective planning which depends on the skills of the planners,

3. Availability of the required labour skills, specific tools, parts and equipment.
4. Proper repair procedures.
5. Adequate time to repair.
6. Verification of repair.

32.6.4 Predictive Maintenance (PDM)

Predictive maintenance (**PDM**)[57] techniques help determine the condition of in-service equipment in order to predict when maintenance should be performed. Every rail infrastructure object has an expected lifespan, which is adjusted after observations of the track. This allows a rail Infrastructure Manager to predict the maintenance date in advance.

This method uses analyses, monitoring, diagnoses and trends. This type of preventive maintenance is carried out each time the value of a given system parameter (nearby) exceeds a predetermined value. The preventive maintenance date can be predicted with the help of these techniques since these can measure the deterioration of the track.

This approach offers cost savings over routine or time-based preventive maintenance, because tasks are performed only when warranted.

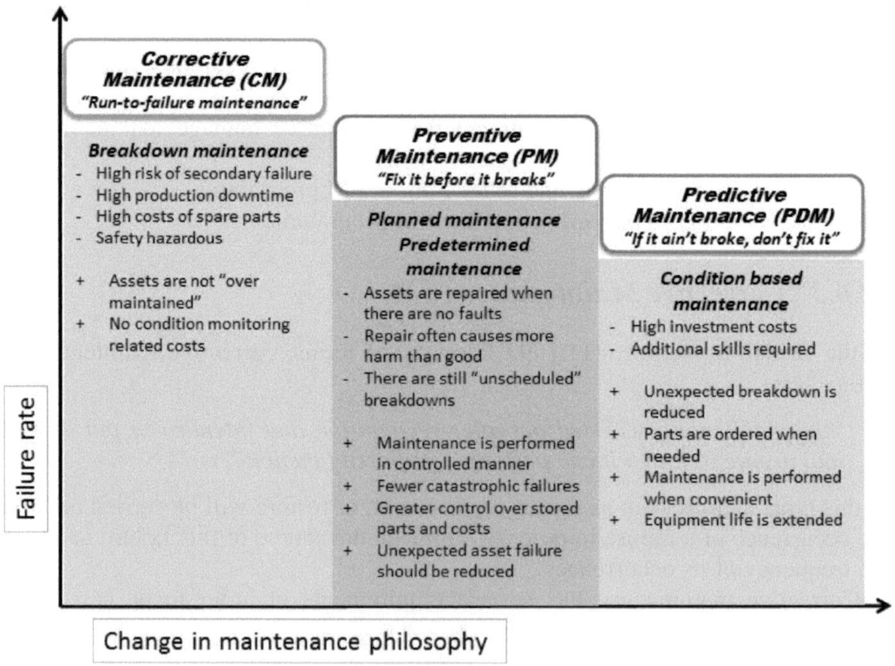

Fig. 142 Changes in maintenance philosophy

[57] Also known as condition-based maintenance.

The main value of predictive maintenance is to allow convenient scheduling of corrective maintenance, and to prevent unexpected equipment failures. The key is **"the right information in the right time"**. By knowing which equipment needs maintenance, the maintenance work can be better planned (spare parts, people etc.). Other values are increased equipment life time, increased plant safety, fewer accidents with negative impacts on environment, an optimised spare parts handling, etc.

The ultimate goal of PDM is to perform maintenance at a scheduled point in time when the maintenance activity is most cost-effective and before the equipment loses performance within a threshold.

Figure 142 presents the changes in maintenance philosophy.

32.7 The Effective (Rational) Maintenance

32.7.1 The "Old Strategy Concept"

We start presenting the "old strategy concept", which is applied till now by several Infrastructure Managers.

Today, maintenance is largely empirical in the planning of maintenance in the short or medium term and executed

- in a predetermined context (e.g. constant maintenance cycles) or
- "extinguishing", i.e. interventions are executed to remove the symptoms of the faults rather than their causes.

The result is the rapid recurrence of faults and constant and costly maintenance cycles. Another disadvantage of this practice is that maintenance is often not necessary (wasting of resources) and additionally in some cases it burdens the asset rather than benefits.

It should be noted that it is uncertain whether the application of such periodic maintenance can secure the reliability of the asset in the long term.

Consequently, there is need for an effective maintenance strategy, which will ensure the required functionality of the track, starting at the design phase of the asset.

32.7.2 Evaluation of Current (Old) Maintenance Strategy

When a railway track is constructed, function and capacity is "purchased", but with it a lifetime of maintenance requirements and the associated cost is acquired. This cost, together with the expected availability and reliability, will play a significant role in the success of the business of the Infrastructure Manager for many years to come [101].

All the elements of the "total approach" come together during the operation in the asset lifecycle. The key questions here are:

- Are we performing all maintenance tasks in a quality manner?
- Are skill levels of the maintenance staff appropriate?
- Are replacement parts available just in time?
- Does our staff have access to the required technical knowledge?
- Are we continually learning, and using this knowledge to improve the maintenance strategies?
- Do all of the above contribute to the expected / desired levels of availability and reliability, with a high level of predictability of cost?
- Are we preserving the integrity and value of the asset base, thereby looking after the capital invested by the investors?

As mentioned, in most of the cases the Infrastructure Managers are applying the old, traditional management strategy.

In order to evaluate the efficiency of this strategy, the following key questions should be answered:

QUESTIONS: Does the current maintenance strategy ensure

- reliable maintenance?
- cost effectiveness of maintenance?
- high track quality for a long time (track durability)?
- that maintenance will be done when really needed and that this will occur as often as it really should?

FINALLY, DOES THE CURRENT STRATEGY ENSURE THE EXECUTION OF A RATIONAL MAINTENANCE?

32.8 The "Future" Maintenance Strategy

The "future" maintenance strategy will be seen (Figure 143[58])

- Rational (reliable, cost effective, assuring long durability, with best use of the available resources – people, equipment and money)
- Based on good track (functional) knowledge and research, especially on specific problems
- Predictive

The following tools will be used to fulfil the above mentioned targets:

(1) RAMS analysis (see SECTION VII)
(2) LCC tools (see SECTION VII)
(3) Track behaviour analysis
(4) Track quality prediction models

---

[58] Based on Prof. Peter Veit (Technical University of Graz) seminars.

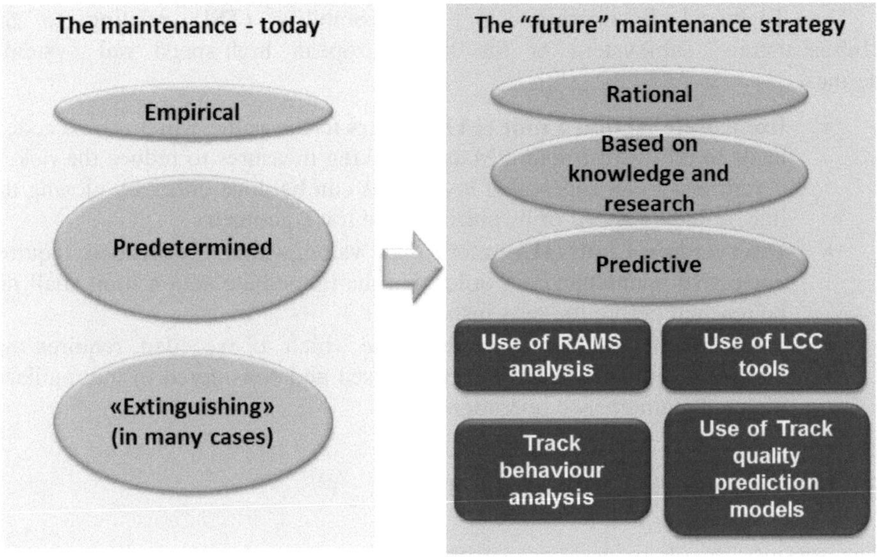

Fig. 143 The transition from the "old" strategy to a "future" maintenance strategy

The transition procedure to a rational maintenance is shown in Figure 144:

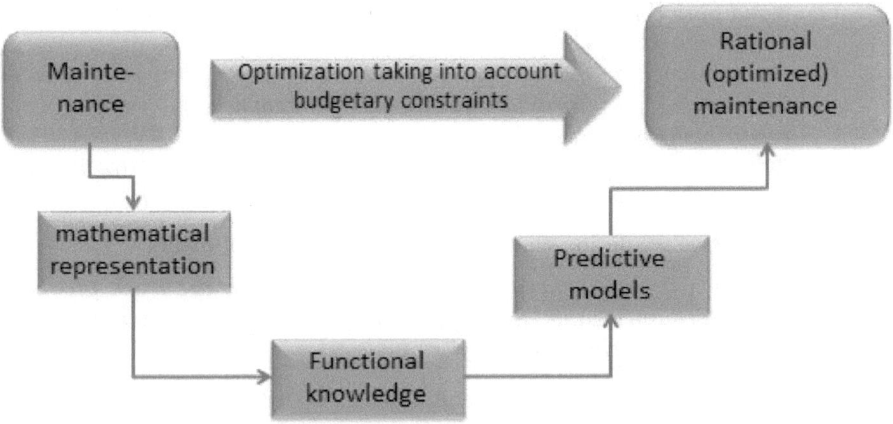

Fig. 144 Transition procedure to a rational maintenance

32.9 The Limit Levels for the Track Defects

Before examining the impact of the different maintenance policies on the service life of the track in dependence with the maintenance interventions and the relevant limits of the track defects, we have to define those limits.

The Technical Specification of Interoperability (TSI) relating to the "Infrastructure" sub-system of the trans-European high-speed rail system[59] defines[60]:

- **Immediate Action Limit (IAL):** refers to the value which, if exceeded, leads to the Infrastructure Manager taking measures to reduce the risk of derailment to an acceptable level. This can be done either by closing the line, reducing speed or by correction of track geometry.
- **Intervention Limit (IL):** refers to the value, which, if exceeded, requires corrective maintenance in order that the immediate action limit shall not be reached before the next inspection;
- **Alert Limit (AL):** refers to the value which, if exceeded, requires that the track geometry condition is analysed and considered in the regularly planned maintenance operations.

32.10 The Service Life of the Track

The service life of an asset is a technical parameter that depends on design, construction quality, operations and maintenance practices, use, and environmental factors.

There are several definitions for service life: the service life of an asset is the

(1) "*total period during which it remains in use, or ready to be used, in a productive process*"[61].

(2) "*the assumed period for which a structure or part of it is to be used for its intended purpose with anticipated maintenance but without major repair being necessary*"[62].

(3) "*a period of time after installation during which a structure (or its parts) meets or exceeds the performance requirements*"[63].

Based on the above we define service life of a track as the following:

Service life of a track is "***the period of time over which the track provides adequate performance and function with anticipated maintenance but without major repair being necessary***".

Listed below are some whole life definitions.

- ***Economic Life.*** The period of time over which an asset will have economic value that is greater than its upkeep and operating costs, or is considered the lowest cost alternative for satisfying a particular need.

[59] EUROPEAN COMMISSION DECISION of 20 December 2007.

[60] The same definitions are given in Euronorm EN 13848-5 (2008): Railway applications. Track. Track geometry quality. Geometric quality levels. Plain line.

[61] Definition by the Organisation for Economic Co-operation and Development (OECD)-Glossary of Statistical Terms.

[62] EN 1990:2002.

[63] See [103].

- ***Design Life.*** The period of time for which an asset is expected to perform according to the technical specifications to which it was produced.
- ***Useful Life.*** The estimated time, in years, that an asset can be expected to serve its intended function if properly constructed in its present application or installation.
- ***Service Life.*** See above.
- ***Physical Life***. The time it takes for an asset to wear out or fail to perform its intended functions because of physical deterioration. The physical life is normally longer than the economic life. Physical life accounts for physical deterioration, not functional or environmental obsolescence.
- ***Technological Life***. The number of years for which replacement parts remain available for repairs and maintenance to an asset after which the asset must be replaced with new technology.
- ***Whole Life.*** The entire life of an asset from the time its components are first manufactured until the time the asset if finally retired from service entirely.
- ***Shelf Life.*** The period of time during which certain Maintenance Stock can be stored and remain suitable for use.

32.11 The Different Maintenance Policies and the Impact on the Service Life of the Track

("Maintenance intervention policies" or "maintenance policies" are here defined as the basic principles by which an Infrastructure Manager is guided in doing maintenance of his assets)

The maintenance policy has a significant influence on the service life of the track. In the following we present in a simplified form the maintenance policies of three imaginary Infrastructure Managers, showing how they influence track performance[64].

32.11.1 Maintenance Policy of INFRASTRUCTURE MANAGER A

After a new track is given to commercial operation, the track deterioration process starts, which means that the defects of the track start to grow.

Infrastructure Manager A intervenes with the appropriate maintenance actions, when the track defects exceed the Alert Limit AL, but before they reach the Intervention Limit IL (Figure 145). After maintenance, the magnitude of the defects is decreased (closed to those of the new track).

The deterioration and maintenance process is continued over time as a cyclic process, but the maintenance intervals become shorter and shorter. When the

[64] Based on Prof. Peter Veit (Technical University of Graz) seminars.

maintenance cycles becomes very short, the maintenance costs increase and the service life of the track is on its end *(the track doesn't provide any more adequate performance and function with anticipated maintenance – see par. 32.10)*, and renewal of the track is necessary.

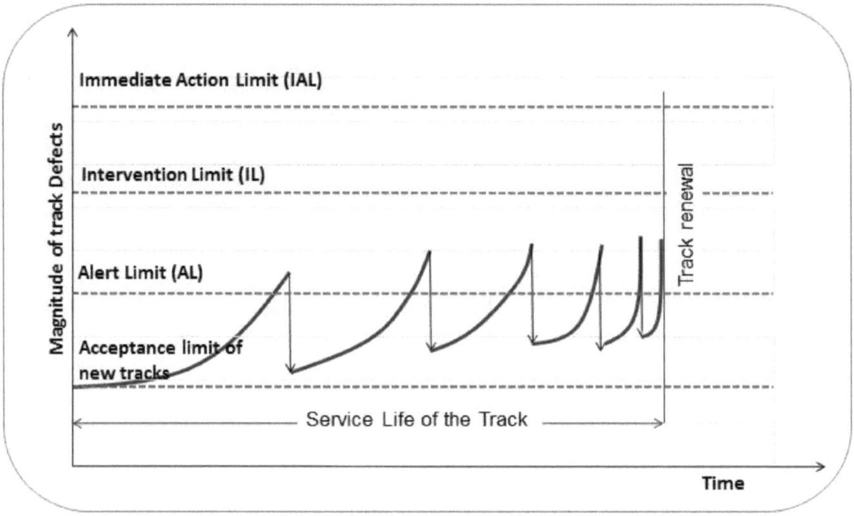

Fig. 145 Maintenance Policy of Infrastructure Manager A

32.11.2 Maintenance Policy of INFRASTRUCTURE MANAGER B

Infrastructure Manager B intervenes with the appropriate maintenance actions, when the track defects exceed the Intervention Limit IL, but before they reach the Immediate Action Limit IAL (Figure 146). After execution of maintenance work, the magnitude of the defects is decreased (their magnitude remains between the AL and IL limit).

The deterioration and maintenance process is continued over time as a cyclic process, but the maintenance intervals become shorter and shorter. When the maintenance becomes very costly, the service life of the track is at its end, and renewal of the track is necessary.

The service life of the track maintained with Maintenance Policy of Infrastructure Manager B is shorter than the respective service life of the track of Infrastructure Manager A.

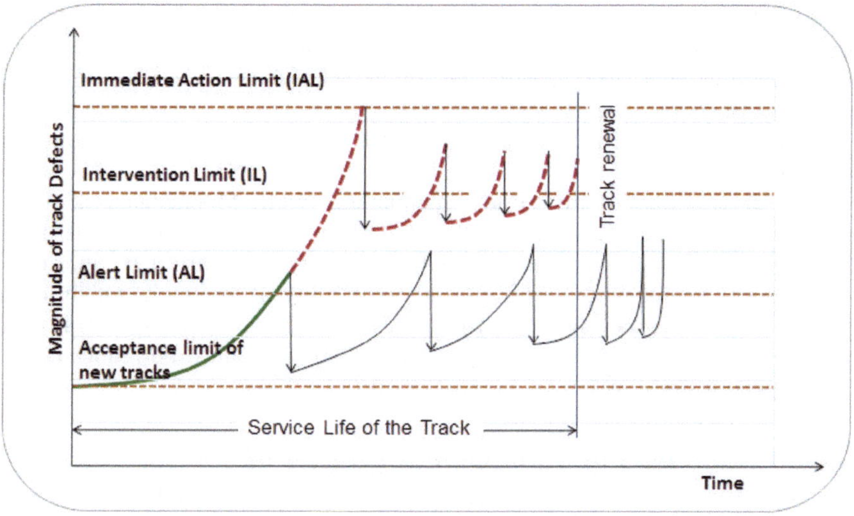

Fig. 146 Maintenance Policy of Infrastructure Manager B

32.11.3 Maintenance Policy of INFRASTRUCTURE MANAGER C

Infrastructure Manager C intervenes with the appropriate maintenance actions, when the track defects reach the Immediate Action Limit IAL (Figure 147). After execution of maintenance work, the magnitude of the defects is decreased (their magnitude remains between the IL and IAL limit).

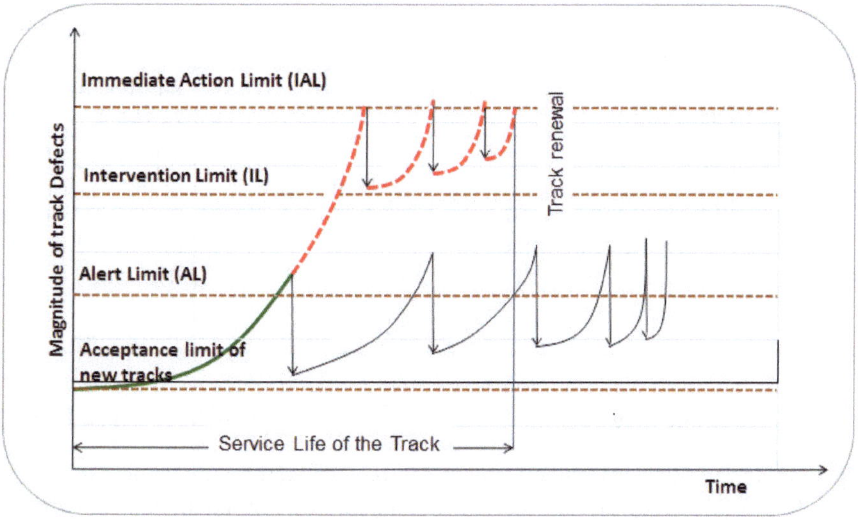

Fig. 147 Maintenance Policy of Infrastructure Manager C

The deterioration and maintenance process is continued over time as a cyclic process, but the maintenance intervals become shorter and shorter. When the maintenance becomes very costly, the service life of the track is at its end, and renewal of the track is necessary.

Track quality is generally not good, but safe due to a high maintenance frequency.

The service life of the track maintained with Maintenance Policy of Infrastructure Manager C is shorter than the respective service life of the track of Infrastructure Managers B and A. The track is very costly to maintain, due to a very frequent maintenance and the limited durability of the track, which leads to the necessity of a renewal of the track earlier than in the case of Infrastructure Manager A and B (in other case the track performance must be reduced, i.e. by setting speed restrictions).

32.11.4 Summary of Maintenance Policies of INFRASTRUCTURE MANAGERS A, B and C

The following table summarizes the maintenance policies of Infrastructure Managers A, B and C:

Infra-structure Manager (IM)	Maintenance Policy	Impact on track quality	Impact on service life of track
IM A	Track is maintained when the magnitude of track defects is between AL and IL	Track quality remains good	The longest possible service life is reached
IM B	Track is maintained when the magnitude of track defects is between IL and IAL	Track quality is moderate	The service life of the track is shorter than that of IM A
IM C	Track is maintained when the magnitude of track defects reach IAL	Track quality is not good, but safe due to a high maintenance frequency	The service life of the track is much shorter than that of IM A. The track is very costly to maintain.

13.12 Cost Savings as a Result of Effective Maintenance Policy

Minimizing the track maintenance cost can be achieved by choosing the optimum time for track renewal, as shown in the following example, showing the maintenance policy of two Infrastructure Managers (IM).

After a new track is given to commercial operation, the track deterioration process starts, which means that the defects of the track start to grow (see par. 32.11) and maintenance works start. The track and its components are aging, so the maintenance costs are growing (cost function C(t) in Figure 148).

At time t_1, Infrastructure Manager B (IM B) decides to renew the track due to the fact that maintenance costs start to increase rapidly. After completion of the works at time t_2, the maintenance costs are almost equal to that of the new track, and maintenance costs start growing from this new start point (following a new cost function $C^*(t)$.

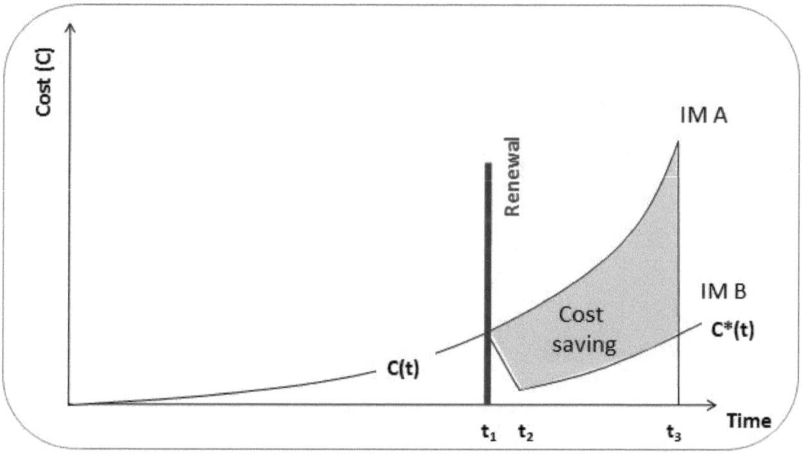

Fig. 148 Choose of the optimum time for track renewal, in order to minimize the maintenance costs

Infrastructure Manager A (IM A) continues to maintain his track, at a high cost way (following cost function C(t)).

Comparing the cost resulting from the maintenance policies of both Infrastructure Managers, we consider a significant cost saving for IM B (dark area in Figure 148).

32.13 Track Maintenance Planning

32.13.1 Introduction

The Infrastructure Manager needs to set up a realistic and efficient maintenance plan, taking into account [20]

- the condition of the current infrastructure,
- assigned track possessions,
- budgets,
- material availability and
- contractor's capacity.

The **objectives** of track maintenance planning can be defined as follows (Figure 149) [18]:

• What is the current condition of the track?	➔	Track **quality** indices
• What will be needed in the short term as well as the long term as far as maintenance is concerned?	➔	**Forecasting** of maintenance actions
• What should be done first?	➔	**Prioritization** of maintenance activities

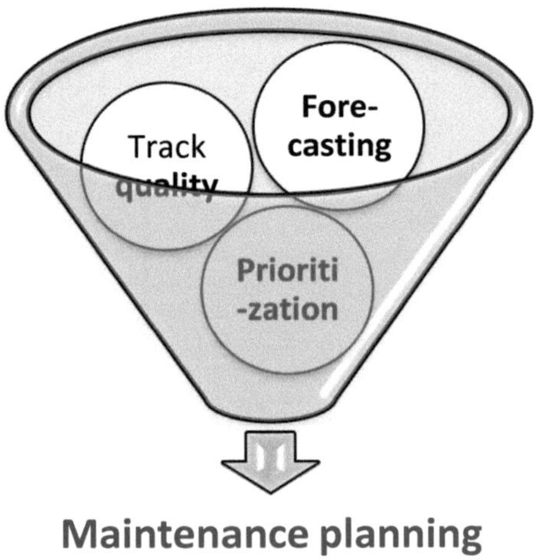

Fig. 149 Parameters for the planning of the track maintenance

RAMS and LCC analysis will play a major role in track maintenance planning, in order to fulfil the above mentioned objectives. A detailed presentation is given in SECTION VII.

A track maintenance optimization is based on a Track Maintenance Management (TMM) system that handles the amount of information needed to make the optimal maintenance decisions over time. The central components of the TMM are

- asset inventory,
- condition measures,
- maintenance input history,
- traffic data,
- cost information.

With this information as well as track deterioration models, the system can be used for selection of maintenance activities.

32.13.2 Track Segmentation

For track maintenance an enormous amount of data is needed. To effectively manage and use all these data in order to analyse them and model the track behaviour, a solution is to divide the linear infrastructure into **homogenous segments with similar track characteristics and similar age and track "history"** (segmentation process) [2].

For the track segmentation following characteristics types are usually taken into account:

- track component types
- layout
- operational characteristics,
- presence of "condition-interfering" objects (e.g. bridges, turnouts, level-crossings, culverts),

An example is shown in Figure 150 [65], where following characteristics are used:

- Traffic load (t/day)
- Radius [m]
- Rail profile
- Sleeper type
- Track construction date
- Last maintenance
- Subgrade type
- Characteristic positions (bridges, turnouts, level-crossings, culverts etc.)

Every single track segment is treated as a separate "organism" with its own unique behaviour characteristics. The behaviour of every such segment can be modelled independently.

The criteria suggested identifying comparable track segments (excluding switches and crossings) are ([2]):

- Type and quality of rail
- Type and quality of fastenings
- Type and quality of sleepers
- Ballast
- Subgrade quality
- Curve radius
- Track gradient
- Installation date of the components.
- Traffic speed
- Axle load of trains / locomotives
- Characteristic positions (bridges, turnouts, level-crossings, culverts etc.)

[65] Based on Prof. Peter Veit (Technical University of Graz) seminars.

- Maintenance history on the analysed segment (operations performed and scheduling):This item should include inspection frequency, rail grinding frequency, ballast cleaning and tamping frequency

Chainage	10	10,1	10,2	10,3	10,4	10,5	10,6	10,7	10,8	10,9	11	11,1	11,2	11,3	11,4
Traffic load (t/day)	30.000	30.000	30.000	30.000	30.000	30.000	30.000	30.000	30.000	30.000	30.000	30.000	30.000	30.000	30.000
Radius [m]	<250	<250	<250	<250	<250	<250	<250	<250	<250	<250	∞	∞	∞	∞	∞
Rail profile	UIC50	UIC50	UIC50	UIC50	UIC50	UIC60	UIC60	UIC60	UIC60	UIC60	UIC60	UIC60	UIC60	UIC60	UIC60
Sleeperf type	WOODEN	WOODEN	WOODEN	WOODEN	WOODEN	BETON	BETON	BETON	BETON	BETON	BETON	BETON	BETON	BETON	BETON
Track construction	1985	1985	1985	1985	1985	1995	1995	1995	1995	1995	1995	1995	1995	1995	1995
Last maintenenace	2006	2006	2006	2006	2006	2006	2006	2006	2006	2011	2011	2011	2011	2011	2011
Subgrade	Type A	Type A	Type A	Type A	Type A	Type A	Type A	Type A	Type A	Type A	Type A	Type A	Type B	Type B	Type B
Characteristic places									Bridge				Turnout		
	SEGMENT 1					SEGMENT 2			SEGMENT 3	SEGMENT 4	SEGMENT 5		SEGMENT 6	SEGMENT 7	

Fig. 150 Example of track segmentation

After the segmentation, additional information (with no significant influence on the segments behaviour), can be assigned to the segments. The different segments will behave in their own way –as different "organisms", which suffer from different "illnesses" (defects/deteriorations) and recover from them by being treated by different "medicine" (maintenance and renewal works) in a different way [2].

Keeping historical data about past behaviour, the segments deterioration behaviour can be modelled, so that the data's can be projected into the future to predict future behaviour (see Chapter 27).

By modelling the future behaviour, the system can define the moment when certain (user-defined) limit values (thresholds) will be reached, signifying the moment when certain maintenance and renewal works (or other) activities should be performed.

32.14 Maintenance Actions

Various types of maintenance actions are presented as encoded in a relevant technical specification of the Hellenic Railways Organization[66]:

100- Superstructure Components

110 - Rails	111	Grinding of new rails
	112	Grinding of rails already in use
	113	Replacement of single rails
	114	Cutting of local depression of the running surface at the rail ends (jointed track)
	115	Restoration of a rail break
	116	Retreading or replacement of welds
120-Fastening system	121	Rearrangement of elastic pads
	122	Replacement of elastic pads
	123	Fastening maintenance
	124	Fastening replacement
130-Sleepers	131	Sleeper replacement
140-Ballast	141	Ballast completions where needed
	142	Ballast cleaning
150-Divers	151	Fixing the position of the anti-creeping and the sleeper anchoring devices
	152	Replacement of the anti-creeping and the sleeper anchoring devices
	153	Replacement of damaged components – materials

[66] Technical Specification E07.01.20 "Maintenance of the track superstructure – General Principles", 2011.

200- Track Devices

210- Switches and crossings	211	Check of the condition (geometry and components)
	212	Checking the safety distances of the switches / crossings and of the their mechanisms
	213	Tightening of the fastenings
	214	Geometry restoration
	215	Replacement of damaged components – materials
220- Expansion joints	221	Inspection of the condition of the expansion joints and respective maintenance
	222	Check and adjustment of the expansion gap
	223	Tightening of the fastenings
	224	Lubrication of the expansion joint

300 – Drainage

301	Cleaning of drains and culverts

400 – Track Geometry

401	Track geometry restoration
402	Restoration of track deformation

500 – Track Laying Characteristics

501	Track gauge rectification
502	Rectification of the verticality of sleepers
503	Rectification of the sleepers spacing
504	Rectification of the joint gaps (jointed track)

600 – Securing the Track Position

601	Inspection and rectification of the points securing the track position

700 – Rail Signs

701	Inspection and rectification of the signs along the track
702	Restoration of the visibility of railway signals

800 – Securing the Loading Gage

801	Removal of obstacles to secure the loading gauge

900 – Level Crossings

901	Maintenance of level crossings
902	Removal of obstacles in the area of level crossings to ensure visibility

1000 – Safety Facilities for Track Personnel and the Public

1001	Inspection and maintenance of handrails, damages to platforms etc.

1100 - Extraordinary Events

1101	Track restoration after extraordinary event (e.g. derailment, flood embankments, slope collapses etc.)

Part D

Tools and Methods for an Effective, High Performance Railway Track

"Don't lower your expectations to meet your performance. Raise your level of performance to meet your expectations. Expect the best of yourself, and then do what is necessary to make it a reality."

Ralph Marston

Section VII: Tools for Rational Maintenance (RAMS and LCC)

"You can't expect to meet the challenges of today with yesterday's tools and expect to be in business tomorrow."

Production is not the application of tools to materials, but logic to work

Peter F. Drucker

In this section tools for the optimization of the track construction regarding technical and economic requirements are presented. LCC and RAMS technology are two acknowledged methods for assisting the optimization process regarding those requirements.

33 Introduction to LCC and RAMS

Key terms:

Optimization process, best technical-operating solution, most economical solution, engineering economics technique, feeling for the cost impact of our decisions, supporting method for professional decision making, quality of service, qualitative and quantitative indicator, continuous process, risk, categories of risk, risk analysis, probability of occurrence of an event, consequence of a hazard, hazard severity levels, "frequency - consequence" matrix, the right product at the right price within the agreed timetable, critical success factors, lowest long term cost of ownership, cost effective solution vs. best solution

Optimization of the track construction or the track components regarding technical and economic requirements is essential for railway companies to fit the market and to compete against other means of transport. LCC[67] and RAMS[68] technology are two acknowledged methods for assisting the **optimization process** regarding those requirements.

Why is the LCC and RAMS analysis important for decision making?

Every day we make **decisions**: we make investments for infrastructure and vehicles, we choose the supplier, we try to stretch the maintenance interval or we even want everything to remain constant. Every one of these decisions is **cost related**. The stakeholders (see Chapter 8) demand today not only for the **best technical-operating** but also for the **most economical solution**.

The current decisions in the railway sector have effects over a long period in the future. So, we **need a feeling for the cost impact of our decisions and a supporting method for professional decision making**. In this regard **LCC and RAMS are the appropriate methods**.

In the case of the optimization of an existing system it is done mainly on the basis of costs. But the costs are related to the investment, the cost for operation, maintenance and non-availability. In case of funding by the government we may have to look for social economics. The relation between technical and economic aspects together with future requirements often makes a traceable assessment difficult.

---

[67] Acronyms for "Life Cycle Cost", see Chapter [36].

[68] Acronyms for "Reliability, Availability, Maintainability, Safety", see Chapter [34].

K. Tzanakakis: *The Railway Track and Its Long Term Behaviour*, STTT 2, pp. 255–257.
DOI: 10.1007/978-3-642-36051-0_33

Targets for the optimizations of the track system are

- Technical and operational performance of components,
- Maintenance procedures,
- Maintenance strategies.

This chapter describes the utilisation of RAMS and LCC methodologies in track maintenance planning.

For the Infrastructure Manager, the key issues for the track maintenance are:

1st Issue: Which is the best way to maintain the track?

2nd Issue: How can we be assured, that over time the current maintenance practice is the most effective one?

33.1 The 1st Issue: Which Is the Best Way to Maintain the Track?

To the 1st issue (which is the best way to maintain the track?) following questions must be answered:

(1) When should the track be maintained?
(2) How can railway traffic be least disturbed?
(3) What maintenance activities should be performed?
(4) How can the hazards be minimized during the execution of maintenance works?

The answer to the above questions leads to the RAMS analysis, as presented in the table below (a deeper analysis follows):

Table 24 Basics of RAMS Analysis

Questions to 1st Issue	Tool	Key word
(1) When should the track be maintained? (Predict "when" to take maintenance actions depending on the failure modes of the track.)	Reliability analysis	**R**eliability
(2) How can railway traffic be least disturbed? (Predict the frequency and duration of track possession periods due to maintenance actions carried out on the track.)	Availability analysis	**A**vailability

Table 24 (*continued*)

(3) What maintenance activities should be performed? (Determine "what" maintenance actions need to be taken on the event of failures occurring on the track and "how" much time is taken to carry out those maintenance actions.)	Maintainability analysis	**M**aintainability
(4) How can the hazards be minimized during maintenance works? (Estimate the risk of carrying out different maintenance actions on the track in terms of severity and cost.)	Safety analysis	**S**afety

Proper RAMS analysis can help in effective maintenance planning of the track infrastructure and meeting maintenance objectives.

33.2 The 2nd Issue: How Can It Be Ensured That over Time the Current Maintenance Practice Is the Most Effective One?

LCC analysis is a method for calculating the total cost of a system or a product over its total lifespan. A very central target is the systematic process for evaluating and quantifying cost impacts. LCC analysis is primary a method for decision making through economic assessment, comparison of alternative strategies and design. Within a LCC analysis all payments - also future payments - will be referred to a reference date using the discount rate (see par. 36.3).

A life cycle cost analysis calculates the cost of a system or product over its entire life span. The method is one of the most recommended for investment projects, assessment of different solutions over the whole life cycle and comparison of various strategy options.

LCC analysis is primary a method for decision making through economic assessment and comparison of alternative strategies and design.

34 The RAMS Analysis

Key terms:

Quality of service, qualitative and quantitative indicator, continuous process, risk analysis, probability of occurrence of an event, consequence of the hazard, hazard severity levels, "frequency - consequence" matrix, risk, categories of risk, the right product at the right price within the agreed timetable, critical success factors, goals for maintenance

34.1 Introduction

Railway RAMS is a major contributor to the **quality of service** provided by the Infrastructure Manager.

A system can be defined as an *"assembly of sub-systems and components, connected together in an organised way, to achieve specified functionality"*. Functionality is assigned to sub-systems and components within a system. The behaviour and state of the system is changed if the sub-system or component functionality changes. A system responds to inputs to produce specified outputs, whilst interacting with an environment.

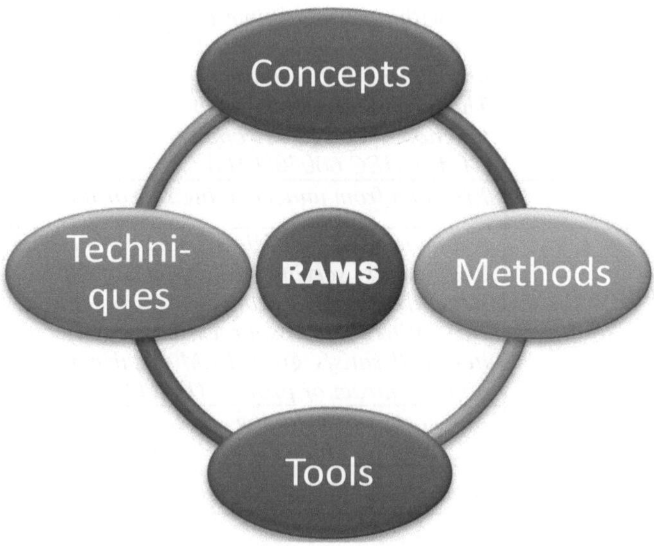

Fig. 151 Constituents to achieve RAMS

K. Tzanakakis: *The Railway Track and Its Long Term Behaviour*, STTT 2, pp. 259–274.
DOI: 10.1007/978-3-642-36051-0_34

RAMS is a characteristic of a system's long term operation and is achieved by the application of established engineering concepts, methods, tools and techniques throughout the lifecycle of the system (Figure 151).

The RAMS of a system can be characterised as a **qualitative and quantitative indicator** of the degree that the system (or the sub-systems and components comprising that system) can be relied upon **to function as specified and to be both available and safe**.

The goal of an Infrastructure Manager is to achieve a defined level of track quality in a given time, safely by execution of cost-effective maintenance actions. Railway RAMS describes the confidence with which the system can guarantee the achievement of this goal.

LCC analysis will help in optimising the cost-effectiveness of maintenance actions derived from RAMS analysis. Cost estimations through LCC help in foreseeing the cost implications of maintenance actions over the service life of the track not just in the short term.

34.2 Definitions

(The definitions are taken from Euronorm EN 50126-1)

Availability:	*The ability of a product to be in a state to perform a required function under given conditions at a given instant of time or over a given time interval assuming that the required external resources are provided.*
Maintainability:	*The probability that a given active maintenance action, for an item under given conditions of use can be carried out within a stated time interval when the maintenance is performed under stated conditions and using stated procedures and resources. (IEC 60050(191))*
Reliability:	*The probability that an item can perform a required function under given conditions for a given time interval (t1, t2). (IEC 60050(191))*
Safety:	*Freedom from unacceptable risk or harm*
RAM programme:	*A documented set of time scheduled activities, resources and events serving to implement the organisational structure, responsibilities, procedures, activities, capabilities and resources that together ensure that an item will satisfy given RAM requirements relevant to a given contract or project. (IEC 60050(191)).*
RAMS:	*An acronym meaning a combination of Reliability, Availability, Maintainability and Safety.*
Hazard:	*A physical situation with a potential for human injury.*

34.3 Key Values for System Operational States

The following key values describing the system operational states have been identified in [24][24.10]:

Reliability	MTBF	Mean Time Between Failure (for corrective maintenance)
	MTBM	Mean Time Between Maintenance for preventive maintenance MTBF and MTTF: Often used to describe equipment or system reliability. Of use when calculating maintenance costs. Meaningful even if the failure rate is not constant.
	Train delaying failures	
Availability	Train delay hours	
Maintainability	MTTR	Mean Time to Repair Often expressed in percentile terms such as “the 95% repair time shall be 1 hour”. This means that only 5% of the repair actions shall exceed 1 hour.
	MDT	Mean Down Time In organizational management, mean down time (MDT) is the average time that a system is non-operational. This includes all time associated with • repair, corrective and preventive maintenance, • self-imposed downtime, and • any logistics or administrative delays. The inclusion of delay times distinguishes mean down time from mean time to repair (MTTR), which includes only downtime specifically attributable to repairs.
	NAMT	Mean Active Maintenance Time
	MMH	Mean Maintenance Hour (MMH)
Safety	Hazard Rate	
	Number of derailment due to asset	
	Number of accidents	

34.4 RAMS Elements

The individual performance indicators are here described. These indicators determine the performance for an asset in terms of reliability, availability, maintainability and safety.

In Systems Engineering, RAMS is providing an **indication of the performance of the designed systems,** or a **quantitative indication of the performance of an asset**. RAMS can be used by the Infrastructure Manager to demand certainty from the contractors that the chosen design solution will meet the requirements over a systems life cycle.

The maintenance decision process is a **continuous process** to make the maintenance decisions effective while meeting the RAMS targets laid for the track system:

- Reliability and maintainability analysis affect the initial maintenance decisions.
- Availability and safety parameters are calculated from reliability and maintainability analysis as well as the maintenance decisions. Availability and safety analysis give feedback to reliability and maintainability analysis and maintenance decisions [111].

34.4.1 Failure

Before presenting the various reliability parameters it is necessary to define the term "*Failure*":

- Failure is "n*on-conformance to some defined performance criterion*" [106]
- Or, failure is "*the termination of the ability to perform the required function*" [115].

34.4.2 Reliability

Reliability is measured by the "*probability that a system or component will work without failure during a specified time interval, with consideration of given operating conditions and environment*".

The failure rate is often described as the **Mean Time Between Failure (MTBF)** (par. 34.3) which determines the time of the average period between failures.

For <u>non-repaired items with constant failure rate</u>, reliability is given by:

$$R = \left(\frac{\text{Number of failures}}{\text{Total operating time}}\right)$$

If failures occur randomly then they can be described by an exponential distribution [118]

$$R(t) = e^{-\lambda \cdot t} = e^{\frac{-t}{MTBF}}$$

Where λ = Failure Rate ($\lambda = 1 / MTBF$)

Example:

For an item with a constant failure rate of one occurrence per operating year and a required time of operation of 11 months, the reliability is given by

$$R(11m) = e^{-\lambda \cdot t} = e^{\frac{-1}{11/12}} = 0{,}39985 \approx 40\%$$

<u>Probability of Failure</u>

The general expression for probability of failure F(t) is given by

$$F(t) = 1 - R(t)$$

The reliability of a certain section of a railway line is for example 95,1%. This means that there is a certainty of 95,1% that traffic can get pass the section of the line on a given time (or 4,9% probability of failure).

Series and parallel operations can be combined into complex block diagrams, as shown in Table 25 (R_S is the system reliability) [114].

Table 25 Reliability Calculations (after [114])

Type branch	Block diagram representation	System reliability[69]
Series	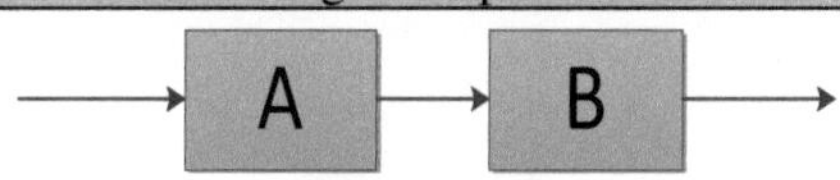	$R_S = R_A \cdot R_B$
Parallel	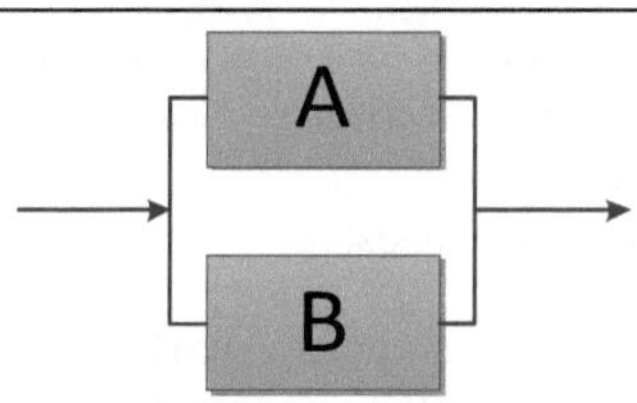	$R_S = 1-(1-R_A)(1-R_B)$
Series - parallel	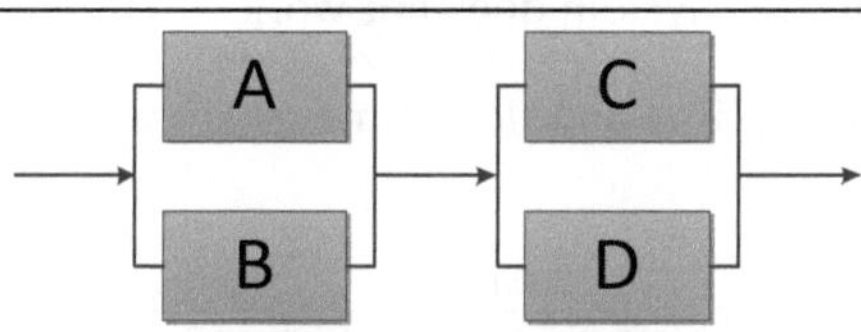	$R_S = [1-(1-R_A)(1-R_B)] * [1-(1-RC)(1-RD)]$
Parallel - series	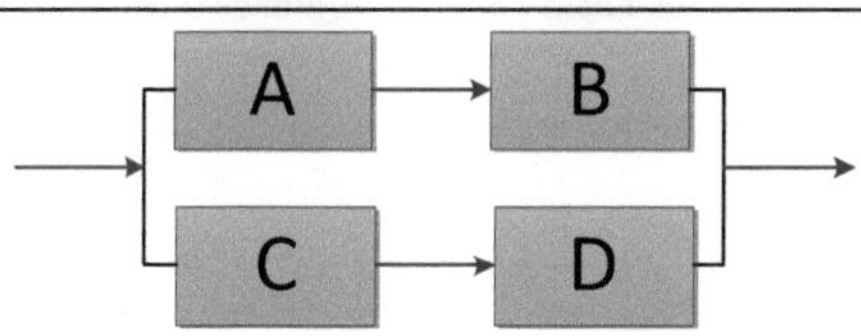	$R_S = 1-[1-(R_A * R_B)] * [1-(R_C * R_D)]$

[69] Assumes all components function independently of each other.

Reliability is obtained from failure rate, as described in par. 34.4.2. Failure rate can be obtained by employing statistical methods[70]. For this to happen, failures must be recorded. In most databases today, failure data sets are scarce or they have a short background. The whole process is summed up in Figure 152.

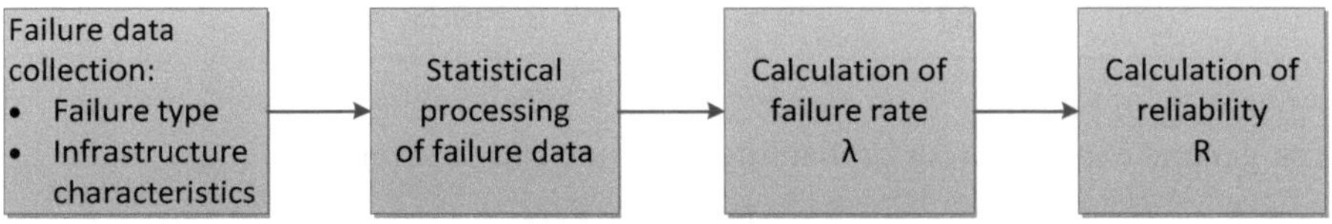

Fig. 152 Reliability and failure rate forecasting procedure [117]

Failure Data Collection

Each Infrastructure Manager has determined the limits for the different failure modes related to track damage.

To provide the data sets with statistical meaning, failure data should be grouped into comparable sets, which has to do with:

- the same failure mode, and;
- comparable track segments (see par. 32.13.2).

34.4.3 Availability

The availability of a component is strongly related to reliability and maintainability. A reliable system requires less maintenance and good maintainability shortens the repair time.

A system has different operational phases:

- Uptime (Up): the time that a system works
- Downtime (Down): the system does not work

Availability is a function of uptime and downtime as shown the following Figure and equation [111].

$$\text{Availability} = \frac{Uptime}{Total\ time} = \frac{\text{Uptime}}{\text{Uptime+Downtime}}$$

[70] Different expressions or cumulative distribution functions can be used for the failure rate from empirical data sets. The expressions or distributions will differ from case to case, and accordingly testing data will need to adapt to the calculation demands. Four different cases are presented in [117]: High number of Failures (constant and non-constant failure rate over component's life (time dependency)) and Low number of failures (constant and non-constant).

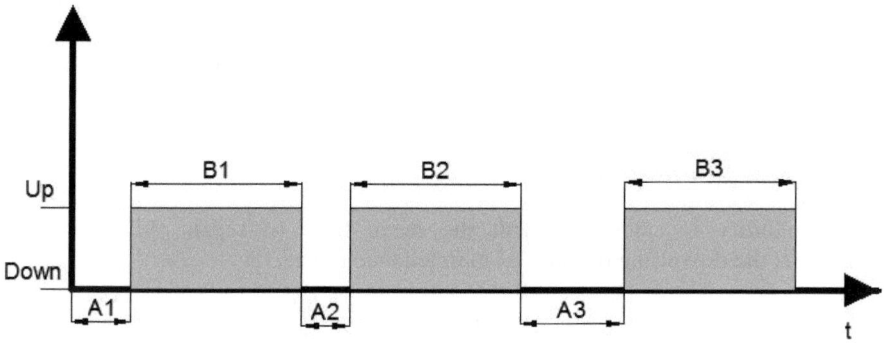

Fig. 153 System Operational States (an example)

There are different ways to determine a systems availability such as

- inherent availability,
- achieved availability and
- operational availability.

Inherent availability for example only takes corrective maintenance into account:

$$Inherent\ availability = \frac{\text{MTBF}}{\text{MTBF} + \text{MTTR}}$$

(MTBF = Mean Time Between Failure)
MTTR = Mean Time to Repair)

Inherent availability is based on an ideal support environment, which assumes that tools, spare parts and personnel are readily available.

Achieved availability is based on the fact that a system needs maintenance and considers preventive and corrective maintenance.

$$\text{Achieved availability} = \frac{\text{MTBM}}{\text{MTBM+ MAMT}}$$

MTBM (Mean Time Between Maintenance
MAMT (Mean Active Maintenance Time)

Achieved availability is also based on an ideal support environment but with consideration of preventive or scheduled maintenance.

Operational availability includes preventive and corrective maintenance, logistics delay time and administrative delay time. It shows the availability in an actual operational environment.

$$\text{Operational availability} = \frac{\text{MTBM}}{\text{MTBM+ MDT}}$$

MTBM (Mean Time Between Maintenance
MDT (Mean Down Time)

MDT (Mean Down Time), includes preventive and corrective maintenance, logistics delay time and administrative delay time (see par. 34.3).

The availability of a track is for example 98,5%. This means that the track should be operational 360 days of 365 days per year (=365 X 98,5%).

34.4.4 Maintainability

Maintainability is calculated with the mean time to repair (MTTR), which represents the downtime because of maintenance done.

$$M_s = 1 - e^{-\mu \cdot t} = 1 - e^{\frac{-t}{MTTR}}$$

where μ is repair (restoration) rate (μ = 1 / MTTR)

34.4.5 Safety

Safety is not easy to measure because of the variety of unsafe situations or accidents. The calculation of the safety component is most often performed by **risk analysis**, where the risks of a specific situation are identified, the occurrence and impact is determined and the total risk is calculated (a deeper analysis follows).

34.5 Risk

> *"the worst thing is to rush into action before the consequences have been properly debated, ... We (Athenians) are capable at the same time of taking risks and of estimating them beforehand"*[71].
>
> Pericles, 430 v. Chr.

34.5.1 Risk Concept

The concept of risk is the combination of two elements:

- the **probability** of occurrence of an event or combination of events leading to a hazard, or the frequency of such occurrences;
- the **consequence** of the hazard.

34.5.2 Failure Categories Suitable for Use in Railway Applications

Table 26 defines failure categories suitable for use in railway applications.

[71] This statement, recorded by Thucydides (460-400 B.C.), was made by Pericles at the public ceremonial burial of the first Athenians killed in the war with the Spartans. Here, Pericles describes the virtues of democracy as practiced by Athens.

Table 26 Failure categories suitable for use in railway applications (after EN 50126.1 [141])

Failure Category	Definition
Significant (Immobilising Failure)	A failure that: • prevents train movement or causes a delay to service greater than a specified time and/or generates a cost greater than a specified level
Major (Service Failure)	A failure that: • must be rectified for the system to achieve its specified performance and • does not cause a delay or cost greater than the minimum threshold specified for a significant failure
Minor	A failure that: • does not prevent a system from achieving its specified performance and • does not meet criteria for Significant or Major failures

34.5.3 Risk Analysis

Risk analysis shall be performed at various phases of the system life cycle by the authority responsible for that phase and shall be documented. The documentation shall contain, as a minimum:

a) analysis methodology;
b) assumptions, limitations and justification of the methodology;
c) hazard identification results;
d) risk estimation results and their confidence levels;
e) results of trade-off studies;
f) data, their sources and confidence levels;
g) references.

Table 27 provides, in qualitative terms, typical categories of probability or frequency of occurrence of a hazardous event and a description of each category for a railway system. The categories, their values, and their numerical scaling to be applied shall be defined by the Infrastructure Manager, appropriate to the application under consideration.

Table 27 Frequency of occurrence of hazardous events (after EN 50126.1 [141])

Category	Description
Frequent	Likely to occur frequency. The hazard will be continually experienced.
Probable	Will occur several times. The hazard can be expected to occur often.
Occasional	Likely to occur several times. The hazard can be expected to occur several times.
Remote	Likely to occur sometime in the system life cycle. The hazard can reasonably be expected to occur.
Improbable	Unlikely to occur but possible. It can be assumed that the hazard may exceptionally occur.
Incredible	Extremely unlikely to occur. It can be assumed that the hazard may not occur.

Consequence analysis shall be used to estimate the likely impact.

Table 28 describes typical **hazard severity levels** and the consequences associated with each severity level for all railway systems. The number of severity levels and the consequences for each severity level to be applied shall be defined by the Infrastructure Manager, appropriate for the application under consideration.

Table 28 Hazard Severity Level (after EN 50126.1 [141])

Severity Level	Consequence to Persons or Environment	Consequence to Service
Catastrophic	Fatalities and/or multiple severe injuries and/or major damage to the environment.	
Critical	Single fatality and/or severe injury and/or significant damage to the environment.	Loss of a major system
Marginal	Minor injury and/or significant threat to the environment	Severe system(s) damage
Insignificant	Possible minor injury	Minor system damage

34.5.4 Risk Evaluation and Acceptance

This paragraph deals with the formation of a **"frequency - consequence" matrix** for evaluation of the results of risk analysis, risk categorisation, actions for risk reduction or elimination of intolerable risks, and for risk acceptance.

Risk evaluation shall be performed by combining the frequency of occurrence of a hazardous event with the severity of its consequence to establish the level of risk generated by the hazardous event. A "frequency - consequence" matrix is shown in Table 29.

Table 29 Frequency - Consequence Matrix

Frequency of occurrence of a hazardous event	**Risk Levels**			
	Severity Levels of Hazard Consequence			
	Insignificant	Marginal	Critical	Catastrophic
Frequent				
Probable				
Occasional				
Remote				
Improbable				
Incredible				

Table 30 defines qualitative **categories of risk** and the actions to be applied against each category. The Infrastructure Manager shall be responsible for defining principle to be adopted and the tolerability level of a risk and the levels that fall into the different risk categories.

Table 30 Qualitative Risk Categories (after EN 50126.1 [141])

Risk Category	**Actions to be applied against each category**
Intolerable	Shall be eliminated
Undesirable	Shall only be accepted when risk reduction is impracticable and with the agreement of the Infrastructure Manager or the Safety Regulatory Authority, as appropriate
Tolerable	Acceptable with adequate control and with the agreement of the Infrastructure Manager
Negligible	Acceptable with / without the agreement of the Infrastructure Manager

Table 31 shows an example of risk evaluation and risk reduction/controls for risk acceptance.

Table 31 Typical Example of Risk Evaluation and Acceptance (after EN 50126.1 [141])

Frequency of occurrence of a hazardous event[72]	**Risk Levels**			
	Severity Levels of Hazard Consequence			
	Insignificant	**Marginal**	**Critical**	**Catastrophic**
Frequent	Undesirable	Intolerable	Intolerable	Intolerable
Probable	Tolerable	Undesirable	Intolerable	Intolerable
Occasional	Tolerable	Undesirable	Undesirable	Intolerable
Remote	Negligible	Tolerable	Undesirable	Undesirable
Improbable	Negligible	Negligible	Tolerable	Tolerable
Incredible	Negligible	Negligible	Negligible	Negligible

Risk Evaluation	**Risk reduction/control**
Intolerable	Shall be eliminated
Undesirable	Shall only be accepted when risk reduction is impracticable and with the agreement of the Infrastructure Manager.
Tolerable	Acceptable with adequate control and the agreement of the Infrastructure Manager
Negligible	Acceptable without any agreement

34.6 Factors Influencing Railway RAMS

The RAMS of a railway system is influenced in three ways:

- by system conditions: sources of failure introduced internally within the system at any phase of the system lifecycle,
- by operating conditions: sources of failure imposed on the system during operation, and

[72] Scaling for the frequency of occurrence of hazardous events will depend on the application under consideration.

- by maintenance conditions: sources of failure imposed on the system during maintenance activities.

These sources of failure can interact.

Table 32 gives some examples on the factors affecting RAMS of the track:

Table 32 Factors affecting track RAMS (after[18])

	Physical parameters	Technical parameters
System conditions	Track curvature (transient curve in, transient curve out, radius)	Quasi-static stress
	Track gradients (start, end, value)	Quasi-static stress
	Rail (rail type, jointed or welded)	Yield strength (Young's modulus)
	Ballast (ballast type, ballast size)	Stiffness, Damping
	Sleeper (sleeper type, sleeper spacing)	Stiffness, Damping, Bending stress
	Fastener (fastener type)	Damping
	Subgrade (geological condition)	Stiffness, Damping
Operating conditions	Track operating conditions: • Loads (annual MGT, maximum axle load) • Environment (temperature)	 Bending stress, Shear stress, Contact stress Thermal stress
	Vehicle operating conditions: • Speed of trains • Vehicle condition (hollow wheels)	 Vertical stress, Lateral stress Dynamic stress
Maintenance conditions	Grinding	Wear rate
	Tamping	Change in track stiffness
	Lubrication	Change in friction co-efficient
	Renewal of track components	Interval of renewal
	Corrective replacements of track components	Failure rate of track components

34.7 The Means to Achieve Railway RAMS Requirements

The means to achieve railway RAMS requirements relates to controlling the factors which influence RAMS throughout the life of the system. Effective control requires the establishment of **mechanisms** and **procedures** to defend against sources of error being introduced during the realisation and support of the system. Such defences need to take account of both **random and systematic failures**.

The means used to achieve RAMS requirements are based on the concept of taking precautions to **minimise the possibility of an impairment** occurring as a result of an error during the lifecycle phases.

Precaution is a combination of:

a) prevention: concerned with lowering the probability of the impairment.
b) protection: concerned with lowering the severity of the consequences of the impairment.

34.8 Methods and Tools for RAMS Analysis

Various tools and methods are used for RAMS analysis. Some of these tools and methods are here presented, for example,

- FMEA (Failure Mode and Effects Analysis),
- FMECA (Failure Mode, Effects and Criticality Analysis),
- FTA (Fault Tree Analysis),
- Failure Block Diagram Analysis,
- CCA (Cause - Consequence Analysis),
- HAZOP (Hazardous Operability Analysis).

Table 33 Methods and tools are used for RAMS analysis

FMEA	**Failure Mode and Effects Analysis (FMEA)** is a systematic tool for identifying: • effects or consequences of a potential product or process failure. • methods to eliminate or reduce the chance of a failure occurring. A successful FMEA activity helps a team to identify potential failure modes based on past experience with similar products or processes, enabling the team to design those failures out of the system with the minimum of effort and resource expenditure, thereby reducing development time and costs.
FMECA	**Failure Mode, Effects and Criticality Analysis (FMECA)** is an extension of FMEA by including a *criticality analysis*, which is used to chart the probability of failure modes against the severity of their consequences. The result highlights failure modes with relatively high probability and severity of consequences, allowing remedial effort to be directed where it will produce the greatest value.
FTA	**Fault Tree Analysis (FTA)** is a diagnostic tool used to predict the most likely failure in a system breakdown. This analysis method is mainly used in the field of safety engineering and reliability engineering to determine the probability of a safety accident or a particular system level (functional) failure.

Table 33 (*continued*)

Failure Block Diagram Analysis	**Failure Block Diagram Analysis** is the analysis of a system based on the component reliabilities. It is a method of modelling how components and sub-system failures combine to cause system failure. Reliability block diagrams may be analysed to predict the availability of a system and determined the critical components from a reliability viewpoint.
CCA	**Cause - Consequence Analysis (CCA)** technique combines cause analysis[73] and consequence analysis[74]. The purpose of CCA is to identify chains of events that can result in undesirable consequences. With the probabilities of the various events in the CCA diagram, the probabilities of the various consequences can be calculated, thus establishing the risk level of the system.
HAZOP	A **Hazard and Operability study (HAZOP)** is a structured and systematic examination of a planned or existing process or operation in order to identify and evaluate problems that may represent risks to personnel or equipment, or prevent efficient operation. A HAZOP is a qualitative technique based on guide-words and is carried out by a multi-disciplinary team (HAZOP team) during a set of meetings, to assess the hazard potential that arise from deviation in design specifications and the consequential effects on the facilities as a whole. This technique is usually performed using a set of guidewords: NO / NOT, MORE / LESS OF, AS WELL AS, PART OF REVERSE, AND OTHER THAN. From these guidewords, scenarios that may result in a hazard or an operational problem are identified. Consider the possible flow problems in a process line, the guide word MORE OF will correspond to high flow rate, while that for LESS THAN, low flow rate. The consequences of the hazard and measures to reduce the frequency with which the hazard will occur are then discussed. This technique had gained wide acceptance in process industries as an effective tool for plant safety and operability improvements.

34.9 System-Effectiveness

System-effectiveness deals with RAMS characteristics of the system. In [109] system-effectiveness is defined as "*the probability that a system may successfully meet an overall operational demand within a given time and when operated under specified conditions*".

[73] Described by fault trees.
[74] Described by event trees.

In short, "***system-effectiveness is the ability of a system to do a job for which it was intended***". It can be defined as a function of the system's operational availability, operational reliability and capability.

Systems characteristic	Symbol	Definition
Operational availability	A_o	*the probability that the track system is operationally available during the train traffic*
Operational reliability	R_o	*probability that during the train traffic operation, the track system will not suffer from any failures*
Capability	C	*the ability that the track system will meet its required objectives*

System-effectiveness (SE) can be calculated as:

System-effectiveness = Operational availability X Operational reliability X Capability

$$SE = A_0 \cdot R_0 \cdot C$$

The higher the system-effectiveness, the better is the track system to achieve its objectives.

In par. 36.2, system effectiveness of various maintenance strategies will be discussed.

35 RAMS in Operation and Maintenance

The system life cycle is an arrangement of phases, each containing tasks, covering the total life of a system from initial concept through to decommissioning and disposal (see Figure 156 in Chapter 36).

The life cycle provides a structure for planning, managing, controlling and monitoring all aspects of a system, including RAMS, as the system progresses through the phases, in order to **deliver the right product at the right price within the agreed timetable**.

A system life cycle, appropriate in the context of railway operation, is shown in Figure 154. The top-down branch (left side) is generally called design and development and is a refining process ending with the manufacturing of system components.

The bottom-up branch (right side) is related to the assembly, the installation, the receipt and then the operation of the whole system.

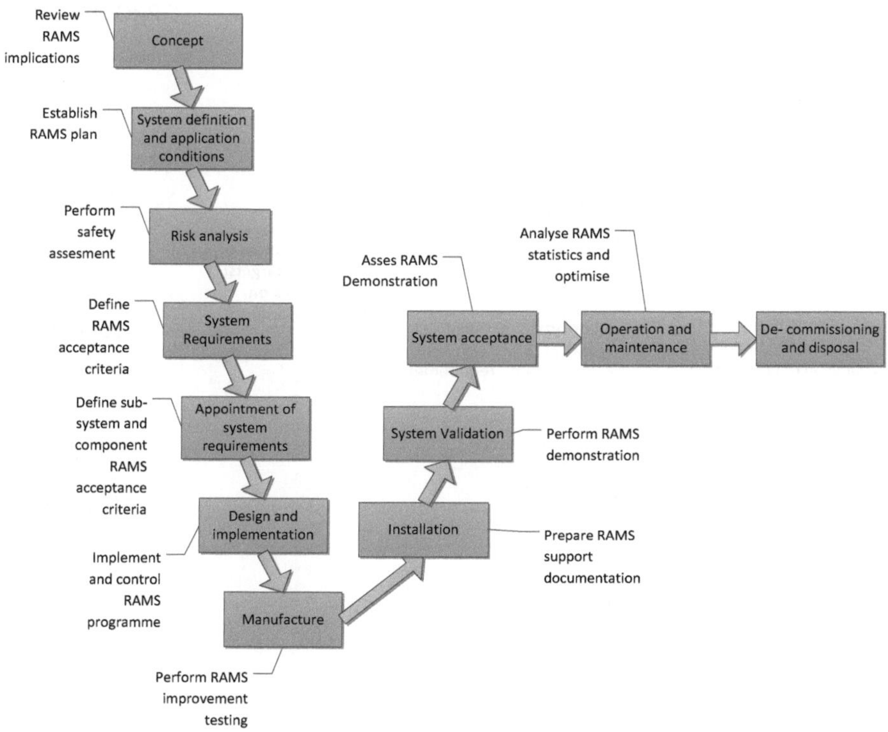

Fig. 154 The "V" representation of RAMS life cycle (after EN 50126.01)

K. Tzanakakis: *The Railway Track and Its Long Term Behaviour*, STTT 2, pp. 275–277.
DOI: 10.1007/978-3-642-36051-0_35

The figure also describes the various RAMS activities being carried out at each phase of the system life cycle (EN 50126.01). To achieve the overall RAMS objectives of the system, it is important to follow systematic RAMS actions throughout the life cycle of the system. As far as RAMS activities are concerned, one of the important phases of the system life cycle is the operation and maintenance phase where RAMS is optimized by analysis of real life failure data.

The objective of this phase is to operate, maintain and support the total combination of components and subsystems such that the compliance with system RAMS requirements is maintained.

Figure 155 illustrates the RAMS process for the track system in the operation and maintenance phase of the system life cycle. It is a continuous improvement process throughout the operation and maintenance phase.

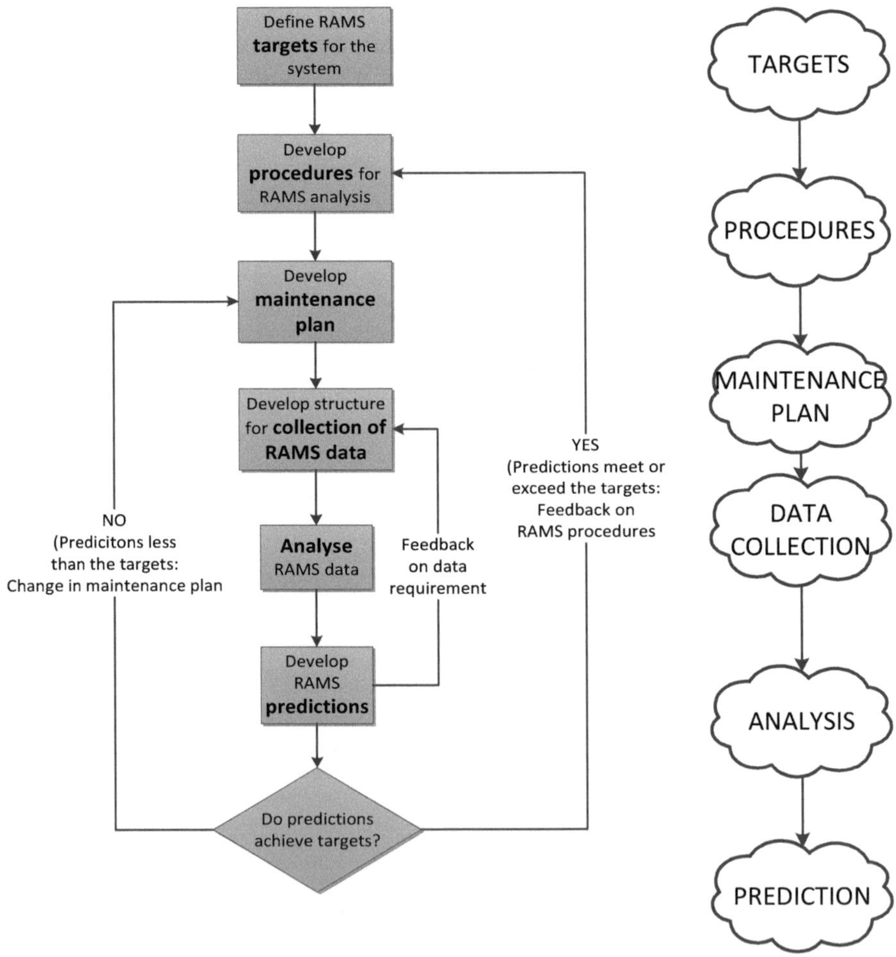

Fig. 155 RAMS process in the operation and maintenance phase (after [18])

RAMS analysis should be done as early as possible in the design stage to ensure the performance calculations are done for the whole system, including all existing subsystems and not only for one subsystem which could be influenced by other subsystems.

36 The LCC Concept

> *Key terms:*
> *long term cost of ownership, engineering economics technique, the cost effective solution vs. the best solution*

"It's unwise to pay too much, but it's foolish to spend too little" *(John Ruston)*

Economical evaluations need to take into account all cost consequences of an investment, independent from the point of time and the place of the payment caused by the investment[75].

Life cycle costing is the process of economic analysis to assess the total cost of acquisition and ownership of an asset.

Several definitions of Life Cycle Cost (hereinafter designated as "LCC") exist:

- A short way of explaining this concept is perhaps to say that the *"life cycle cost of an item is the sum of all funds expended in support of the item from its conception and fabrication through its operation to the end of the useful life"*.
- ISO 15686-5[76] defines LCC *"the cost of an asset, or its parts throughout its life cycle, while fulfilling the performance requirements"*.

The LCC of a physical asset begins when its acquisition is first considered, and ends when it is finally taken out of service for disposal or redeployment (Figure 156).

The method of life cycle costing has been widely accepted in the field of transport during the last years, as a professional cost prognosis method.

Life cycle costing is another term and it is defined in ISO 15686-5 as "*methodology for systematic economic evaluation of life cycle costs over a period of analysis as defined in the agreed scope*".

LCC asks these simple questions:

- What do I need now and how much will it cost me?
- What will I need to do in the future because I have done it and how much will that cost me?
- How long is the "future"?
- How do I evaluate future costs vs. current costs?

Or

- How much does it cost?
- How much does it cost to run?

---

[75] Veit. P., (2010), Workshop LCC Railway Infrastructure, Athens.

[76] ISO 15686-5 Buildings & Constructed Assets: Service Life Planning: Life Cycle Costing.

K. Tzanakakis: *The Railway Track and Its Long Term Behaviour*, STTT 2, pp. 279–292.
DOI: 10.1007/978-3-642-36051-0_36

An appropriate cost analysis provides important inputs in the decision making process with regard to product (here: railway track) design, development and use, evaluating operating and maintenance strategies to optimise long term investment.

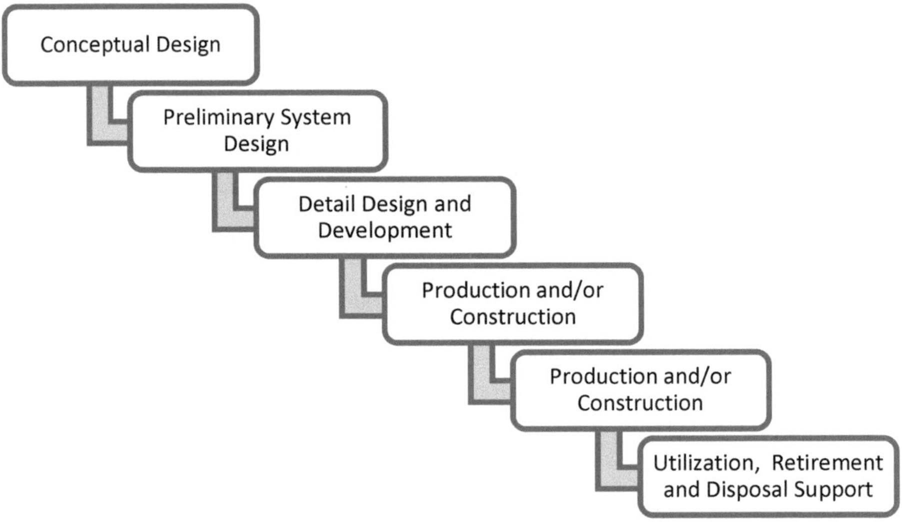

Fig. 156 System Life Cycle Phases[77]

36.1 Why Use LCC?

LCC gives emphasis on enhancing economic competitiveness by working for **the lowest long term cost of ownership** which is not an easy answer to obtain. There are typical problems and conflicts observed in most companies, when different groups of interest in the company have different criteria for taking decisions [108]:

Table 34 Typical problems and conflicts in most companies

Group of Interest….	…. and the only criteria considering
Project Engineering	minimize capital costs
Maintenance Engineering	minimize repair hours
Production	maximize uptime hours
Reliability Engineering	avoid failures
Accounting	maximize project net present value
Shareholders	increase stockholder wealth

[77] Blanchard B., Logistics Engineering & Management (6th Edition), Prentice Hall.

Management is responsible for harmonizing these potential conflicts under the banner of operating for the lowest long term cost of ownership.

LCC can be used as a management decision tool for harmonizing the never ending conflicts between the various groups of interest, by focusing on facts, money, and time.

LCC is important for engineers to help think like MBAs and act like engineers! [108]

In this respect, LCC can be seen as an **engineering economics technique**.

36.2 LCC in Maintenance Optimisation

1st step: LCC calculation

Figure 157 shows how RAMS affects LCC calculation at various stages. The maintenance strategy[78] (MS) with lowest LCC is considered as the cost effective solution to be implemented in the infrastructure operations.

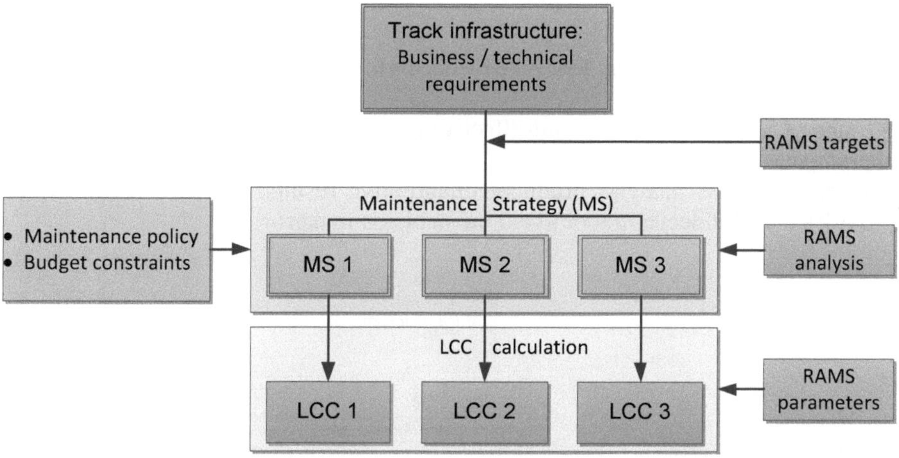

Fig. 157 Maintenance decisions based on LCC [18]

2nd step: Calculation of the system-effectiveness analysis of different maintenance alternatives

As referred above, the maintenance strategy[79] (MS) with lowest LCC is considered as the cost effective solution to be implemented in the infrastructure operations[80].

[78] The maintenance strategy can be a single maintenance action (e.g. grinding) or a cluster of maintenance actions.

[79] The maintenance strategy can be a single maintenance action (e.g. grinding) or a cluster of maintenance actions.

[80] As presented in 36.2.

But **it isn't obvious that the cost effective solution is the best solution**, due to the fact that every solution has a different **system effectiveness** (see par. 34.9).

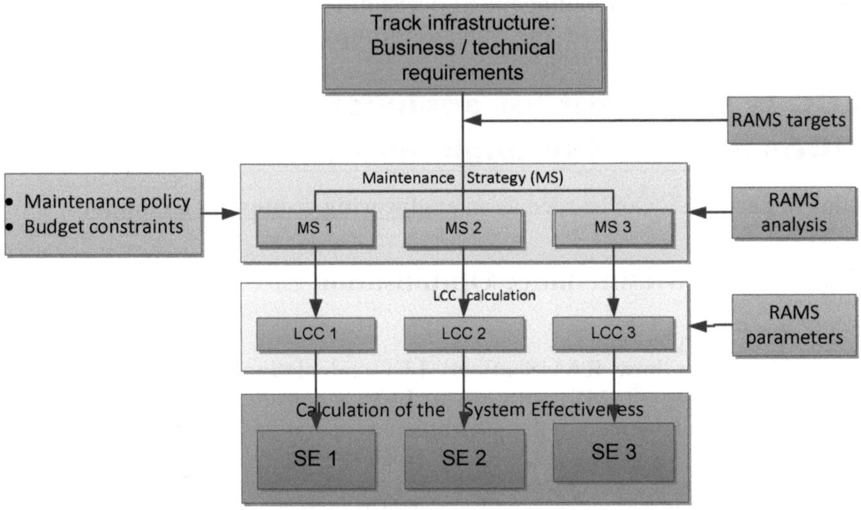

Fig. 158 System-effectiveness analysis

3rd step: Calculation of the cost-effectiveness analysis of different maintenance alternatives

Cost-effectiveness analysis produces quantitative results with risk analysis, and provides a useful decision tool to aid the decision maker.

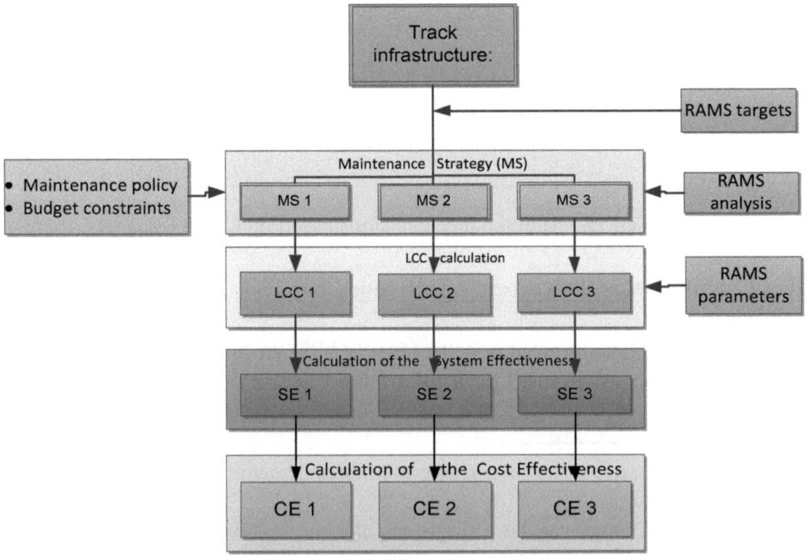

Fig. 159 Cost-effectiveness analysis

Cost-effectiveness (CE) of a maintenance alternative "x" is the quotient of LCC and System-Effectiveness (SE) of the relevant maintenance alternative:

$$CE_X = \frac{LCC_X}{SE_X} = \frac{LCC_X}{A_{0X} \cdot R_{0X} \cdot C_X}$$

Where

A_{0X} the Operational availability (par. 34.9)
R_{0X} the Operational reliability (par. 34.9)
C_{0X} the Capability (par. 34.9)

Table 35 shows the calculation of cost-effectiveness from the LCC values of different maintenance alternatives. The higher the cost effectiveness, the better the maintenance alternative is.

Table 35 Calculation of cost-effectiveness of maintenance alternatives

Maintenance alternatives	Life Cycle Cost (LCC)	System effectiveness	Cost effectiveness CE = SE/LCC
MS 1	LCC 1	SE1	CE1 = SE1 / LCC 1
MS 2	LCC 2	SE2	CE2 = SE2 / LCC 2
MS 3	LCC 3	SE3	CE3 = SE3 / LCC 3

As LCC, by definition, deals with the future and the future is unknown, a risk analysis should be carried out after the performed calculation.

36.3 Economic Calculations for LCC

How do you compare future costs with current costs?

The literature shows a broad variation of economic evaluation methods for LCC analysis. They all have their advantages and disadvantages. The methods have been formed for different purposes and the user should be aware of their limitations. Table 37 [110] illustrates the six main economic evaluation methods for LCC, their advantages and disadvantages and for what purposes they can be used.

Engineers always want a simple, single value, criteria for a project—the answer for LCC is called Net Present Value (NPV):

The NPV of project is the most useful and one of the most commonly used criteria for determining whether a project should be accepted. The net present value formula is:

$$NPV = \sum_{t=0}^{n} \frac{(B_t - C_t)}{(1+r)^t}$$

Where,

B_t are project benefits in period t
C_t are project costs in period t
r is the appropriate financial or economic discount rate
n is the number of years for which the project will operate

In the following the NPV method will be more detailed presented.

For input to the LCC calculation, future costs are converted to their current equivalent by using a suitable discount rate: in project analysis, any costs and benefits of a project that are received in future periods are discounted, or deflated by some factor, r. The factor used to discount future costs and benefits is called the discount rate and is usually expressed as a percentage.

Example:

100 Euro receivable today is more than 100 Euro receivable a year later, as 100 Euro received today will earn interest or profits and shall accumulate to more than 100 Euro in a year's time. Alternatively, 100 Euro received today can be used to reduce borrowing thereby avoiding interest payments as well as reducing debts by 100 Euro. Assuming that the Railways' cost of finance is its current dividend rate (say 6% per year), Euro 106 received a year hence should be worth 100 Euro today and 100 Euro which may be received in a year's time is worth about 94 Euro today (actually it is worth 94.34 Euro). Likewise, the present value of 100 Euro receivable 2 years hence is about 89 Euro, and so on. In this way the cash flow for the project in any future year can be discounted to obtain the present value.

For example, suppose the project is expected to yield a stream of benefits equal to $B_0, B_1, B_2, \ldots B_n$ and to incur a stream of costs equal to $C_0, C_1, C_2, \ldots C_n$ in years 0, 1,2, . . . n. Then in each period the net benefits (benefits minus costs) of the project will be:

$$(B_0 - C_0), (B_1 - C_1), (B_2 - C_2), \ldots (B_n - C_n)$$

This is simply the project's net benefit flow.

Assuming that the discount rate, r, is constant, then the discounted cash flow of the project can be represented as:

Year	Benefit flow
Year 0	$B_0 - C_0$
Year 1	$\frac{(B_1 - C_1)}{(1+r)}$
Year 2	$\frac{(B_1 - C_1)}{(1+r)^2}$
Year 3	$\frac{(B_1 - C_1)}{(1+r)^3}$
Year n	$\frac{(B_1 - C_1)}{(1+r)^n}$

Once future net income streams have been discounted in this way, expenditures and revenues from all the different time periods will be valued in units of similar value - present day units of currency. They will then be directly comparable with each other and can be added together. Adding the discounted net benefits from each year of the project's life, its discounted net benefit flow, gives a single monetary value called the project's net present value, NPV. For, the previous example, the project's NPV is:

Table 36 Example: Cash flow of a railway project - discounted at 6% discount rate (million Euro)

Year (t)	**Costs** (1)	**Benefits (B)** (2)	**Net Benefits** (3)=(2)-(1)	**Discount Factor $1/(1+r)^t$** (4)	**Net Benefits** (5)=(3)*(4)
0	100	0	-100	1	-100,00
1	400	50	-350	0,943	-330,19
2	200	150	-50	0,890	-44,50
3	100	200	100	0,840	83,96
4	100	200	100	0,792	79,21
5	100	200	100	0,747	74,73

Table 36 (*continued*)

6	100	200	100	0,705	70,50
7	100	200	100	0,665	66,51
8	100	350	250	0,627	156,85
Total	**1.300**	**1.550**	**250**	**NPV =**	**57,06**

The net present value criterion of a project is the single most important measure of the project's worth.

In the above table, an r = 6% discount rate is used to discount the net benefits of a railway project. The project's NPV can then be estimated by just adding up these discounted net benefits. Columns (1), (2) and (3) show the non - discounted costs, benefits and net benefits (benefits-costs) of the railway project.

Column (4) gives the discount factor, $1/(1+r)^t$, by which the non-discounted net benefits in column (3) are multiplied, to obtain the discounted value of these net benefits in each year, t, shown in column (5). These discounted net benefits can then be added together to obtain the total discounted net benefits, or net present value, of the project.

The bottom line of the table shows that the NPV comes to 57,06 million Euro if an 6% discount rate is used. A NPV greater than zero indicates that the discounted benefits of the project are expected to be greater than its discounted costs and the project will therefore be worth undertaking.

This example illustrates how crucially the estimation of a project's NPV depends on the discount rate employed[81]. A lower discount rate would have deflated future income by less and increased NPV of the project. A higher discount rate would have deflated future income more heavily and decreased the NPV of the project, possibly changing it from positive to negative. The selection of the appropriate discount rate is therefore a very important issue in project appraisal.

In the above example, a discount rate of r=7% gives a NPV of 32,85 Million Euro, and a discount rate of r = 8% gives a NPV of 10,44 Million Euro. A discount rate of r = 8,5% gives a negative NPV (-0,15 Million Euro).

[81] The discount rate is roughly the opportunity cost of capital: it is the cost of using the capital in one project renouncing to earn a return in another project. Its value is defined mostly empirically for a given project, in a given country or region, for a given firm and at a given time. The value of the discount rate can have a very serious impact on the decision making process of a cost benefit or life cycle cost analysis ([24][24.11]).

(A deeper understanding can be achieved by studying as an example the extensive analysis for the Kansas City Light Rail Project[82]).

The NPV of a project is –as presented- sum of the present values of the net cash flows for all the years of the project's economic life (present value of incomes minus present value of expenses). Projects and processes with the greatest NPV is usually the winner. Often incremental changes on a project, can lead to a positive NPV. Thus many improvement projects must be selected on the least negative NPV values from many alternatives.

NPV in decision making:

If….	**It means ….**	**Then ….**
NPV > 0	the investment would add value to the Infrastructure Manager	the project may be accepted
NPV < 0	the investment would subtract value from the Infrastructure Manager's or government's available budget	the project should be rejected
NPV = 0	the investment would neither gain nor lose value for the Infrastructure Manager's or government's available budget	We should be indifferent in the decision whether to accept or reject the project. This project adds no monetary value. Decision should be based on other criteria, e.g. strategic positioning or other factors not explicitly included in the calculation.

Engineers must be concerned with life cycle costs for making important economic decisions through engineering actions. Management deplores engineers who are engineering bright but economics dim [108].

Cost Considerations

Values for the costs should be as accurate as possible. Greater effort may be required for the most significant cost variables. Values can be derived from:

- a direct estimation from known costs;
- historical data (i.e. from similar track construction works);
- models based on expected performance, averages etc.;
- best guesses of future trends in technology, etc.

For each cost, there should be an associated time profile of when the cost occurs (or recurs).

[82] Sudhakar Raju, (2008), *Project NPV, Positive Externalities, Social Cost-Benefit Analysis - The Kansas City Light Rail Project*, Rockhurst University, `http://www.nctr.usf.edu/jpt/pdf/JPT11-4Raju.pdf`

Table 37 The advantages and disadvantages of economic evaluation methods for LCC (after [110])

Method	What does it calculate	Advantage	Disadvantage	Usable for
Simple payback	Calculates the time required to return the initial investment. The investment with the shortest payback time is the most profitable one.	Quick and easy calculation. Result easy to interpret.	Does not take inflation, interest or cash flow into account.	Rough estimation to see if the investment is profitable (Flanagan et al., 1989).
Discount payback method (DPP)	Basically the same as the simple payback method, it just takes the time value into account.	Takes the time value of money into account.	Ignores all cash flow outside the payback period	Should only be used as a screening device, not as a decision advice .
Net present value (NPV)	NPV is the result of the application of discount factors, based on a required rate of return to each years projected cash flow, both in and out, so that the cash flows are discounted to present value. In general, if the NPV is positive it is worthwhile investing. But as the focus in LCC is on cost rather than on income, the usual practice is to treat cost as positive and income as negative. Consequently, the best choice between two competing alternatives is the one with minimum NPV.	Takes the time value of money into account. Generates the return equal to the market rate of interest. It uses all available data.	Not usable when the comparing alternatives have different life lengths. Not easy to interpret.	Most LCC models utilize the NPV method . Not usable if the alternatives have different life lengths .

Table 37 (*continued*)

Method	What does it calculate	Advantage	Disadvantage	Usable for
Equivalent annual cost (ECA)	This method expresses the one time NPV of an alternative as a uniform equivalent annual cost. For that it takes the factor present worth of annuity into account .	Different alternatives with different life lengths can be compared (ISO, 2004).	Just gives an average number. It does not indicate the actual cost during each year of the LCC (ISO, 2004).	Comparing different alternatives with different life lengths (ISO, 2004).
Internal rate of return (IRR)	The IRR is a discounted cash flow criterion which determines an average rate of return by reference to the condition that the values be reduced to zero at the initial point of time (Moles and Terry, 1997). It is possible to calculate the test discount rate that will generate an NPV of zero. The alternative with the highest IRR is the best alternative (ISO, 2004).	Result gets presented in percentage which gives an obvious interpretation.	Calculations need a trial and error procedure. IRR can only be calculated if the investments will generate an income .	Can only be used if the investments will generate an income, which is not always the case in the construction industry .
Net saving (NS)	The NS is calculated as the difference between the present worth of the income generated by an investment and the amounted invested. The alternative with the highest net saving is the best.	Easily understood investment appraisal technique.	NS can only be used if the investment generates an income.	Can be used to compare investment options (ISO, 2004). But only if the investment generates an income .l

36.4 LCC Process

The estimation of the life cycle cost of the track can be executed in different steps, as shown in Figure 160. The input parameters for each step in the LCC process and the corresponding outputs are described in the figure and the table below.

Table 38 LCC process (after [18])

Process Step	Process	Description
P1	Start of LCC analysis of track section	Understanding of the technical characteristics of a track section as well as the utilisation of the track in terms of tonnage and frequency of trains.
P2	Defining per unit length of track	Determination of the track to compare the cost data of different track configurations in the best possible way (see par. 32.13.2).
P3	Estimating track deterioration	Track deterioration depends both on various track as well as vehicle characteristics. Estimation of total track failures will be done by means of track failure data and degradation models of track and vehicle.
P4	Estimating track maintenance volume	Track maintenance volume consists of all the corrective and preventive maintenance as well as renewal activities. This is estimated by means of RAMS analysis of the various failure modes.
P5	Estimating track possession time	A reference timetable describes the type of traffic, frequency of trains as well as the train operational hours per day. Track possession time determines the track availability, train speed restriction hours and train delay by means of track failure modes and maintenance cost. Track possession times can be calculated based on the corrective and preventive maintenance actions.
P6	Estimating derailment probabilities	Train derailment probabilities are estimated by track failure modes, maintenance volume and track possession time.

Table 38 (*continued*)

Process Step	Process	Description
P7	Estimating total cost (LCC)	LCC is calculated by considering all the costs in the life cycle phases as well as the consequential costs.
P8	Estimating risk in LCC	Uncertainty analysis is done to estimate the variable costs in LCC by simulating the several risky RAMS variables in life cycle cost estimates.
P9	Estimating Net Present Value (NPV)	Net present value (NPV) can be calculated by means of discounting rate.

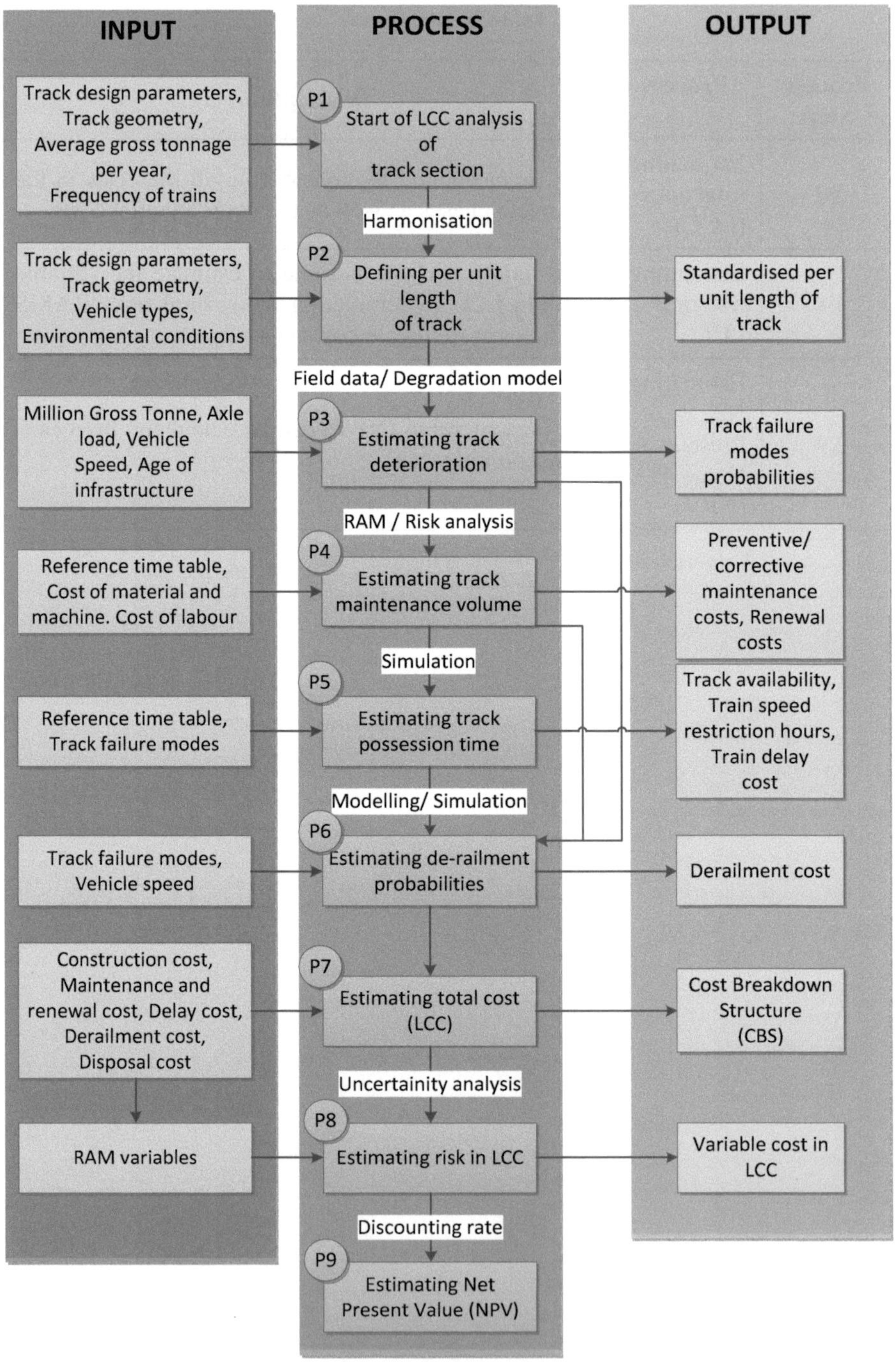

Fig. 160 Process for LCC analysis in the operation and maintenance phase (after [18])

37 Maintenance Planning Using LCC KAI RAMS

The maintenance strategy consists usually of various **critical success factors** that are necessary to achieve the overall **goals for maintenance**.

Effective measurement of the condition of the track is necessary for the achievement of track maintenance goals. Figure 161 describes an approach to achieve track maintenance goals by analysing track condition data using RAMS and LCC (input for the LCC analysis see par. 36.4).

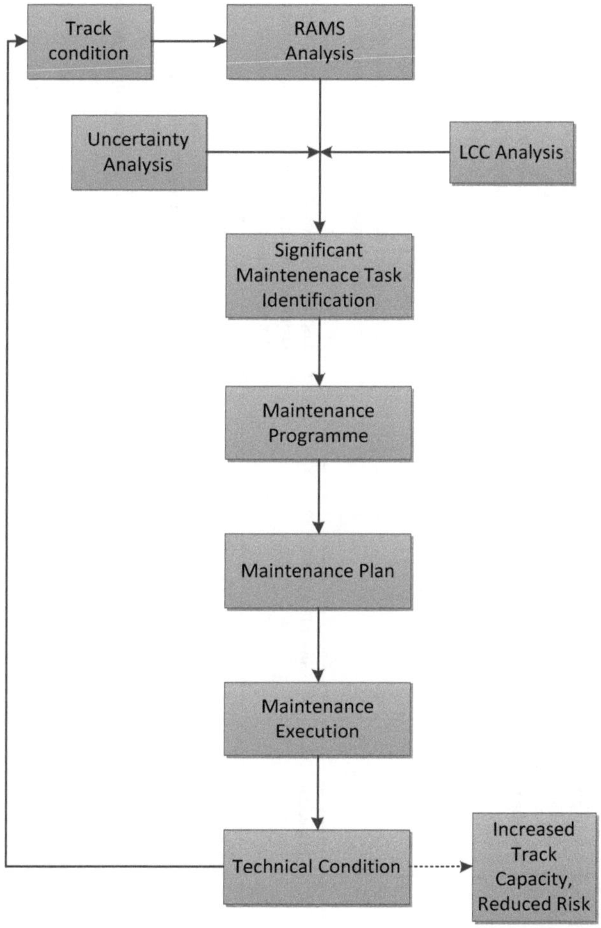

Fig. 161 Maintenance planning using LCC και RAMS (after [18])

K. Tzanakakis: *The Railway Track and Its Long Term Behaviour*, STTT 2, pp. 293–296.
DOI: 10.1007/978-3-642-36051-0_37

Table 39 Reliability analysis in track maintenance planning (after [18])

Track component	Type of maintenance	Maintenance decisions	Reliability analysis	RAMS database
Rail	Grinding	• When to grind?	• Grinding interval is determined by developing a trend of RCF defects data of rail. The point of grinding is determined when the trend reaches the number of defects fixed for the rail	• Type and number of RCF defects. • Past grinding dates.
Track	Tamping	• When to tamp?	• Tamping interval is determined by developing a trend of the standard deviations measured from the desired dimensions of track geometry. The point of tamping is when the trend reaches the standard deviations fixed for the track.	• Track stiffness. • Track geometrical quality indices. • Past tamping dates.
Rail	Inspection	• How often to inspect?	• Inspection interval is determined by analysing the rail car inspection data and developing a trend of number of rail defects formed in a specific interval of time	• Rail car inspection data. • Visual inspection data • Hand held equipment data
Ballast	Cleaning	• When to clean ballast	• Ballast cleaning life estimation *Forecasting of ballast cleaning life is done by developing a probability distribution of past ballast cleaning data and ballast condition factor.* *Ballast condition factor is determined by the percentage of void in the ballast.*	• Past ballast cleaning dates • Condition of ballast.
Rail	Renewal	• When to renew rail in a given track section • What percentage of the rail need to be renewed in a given track section?	• Rail service life estimation for different curve radii. *Rail service life is determined from rail fatigue life and rail wear life whichever comes earlier. Forecasting of rail fatigue life is done by limiting number of RC'F detects per kilometer of track per MGT whereas rail wear life is estimated by keeping a threshold value for rail wear Residual rail life is calculated by using the current condition of rail.*	• Historical rail installation dates and age of rail when installed. • Type and number of RCF defects. • Rail wear profile.

Table 39 (*continued*)

Track component	Type of maintenance	Maintenance decisions	Reliability analysis	RAMS database
Sleeper	Renewal	• When to renew sleepers in a given track section? • How many sleepers need to be renewed in a given period of time in a given track section?	• Sleeper service life estimation for different curse radii. *Forecasting of sleeper service life is done by developing a probability distribution of past sleeper renewal data* *Remaining useful life of sleeper is calculated by using the current conditions of sleepers.*	• Historical sleepers installed dates and conditions when installed • Sleepers condition distribution (e.g.good, medium and bad) in a given track segment. The segmentation is based on the failure modes on the sleepers. • Used life of good, medium and bad sleepers is in terms of percentage of average sleeper life.
Ballast	Renewal	• When to renew ballast	• Ballast service life estimation for different curve radii. *Forecasting of ballast life is done by developing a probability distribution of past ballast renewal data.*	• Historical ballast installation dates • Condition of ballast
Rail	Replacement	• When to replace the rail	• Rail break life estimation. Rail break life forecasting is done by developing a probability distribution of past rail breaks data. It can be also estimated by identifying the rail defects which have got the potential for rail breaks and developing a trend of those defects forming the rail breaks.	• Historical rail break data • Rail defects data • Wear limit.

Track maintenance task identification includes: (see also Chapter 33).

- when the infrastructure needs to be maintained,
- what maintenance actions need to be carried out (preventive or corrective),
- which maintenance action will meet the track availability objective,
- what maintenance action will secure the safety of the system, etc.

38 Integration of RAMS and LCC in the Project Design Process

There are a number of conceptual models for the integration of RAMS and LCC in the design process. In this section a model is described to show how RAMS and LCC can be implemented in sectors such as the production industry.

This model described by Ogink & Al-Jibouri [113] is based on the standard design process from Systems Engineering and fits strong with the construction industry (Figure 162) [111].

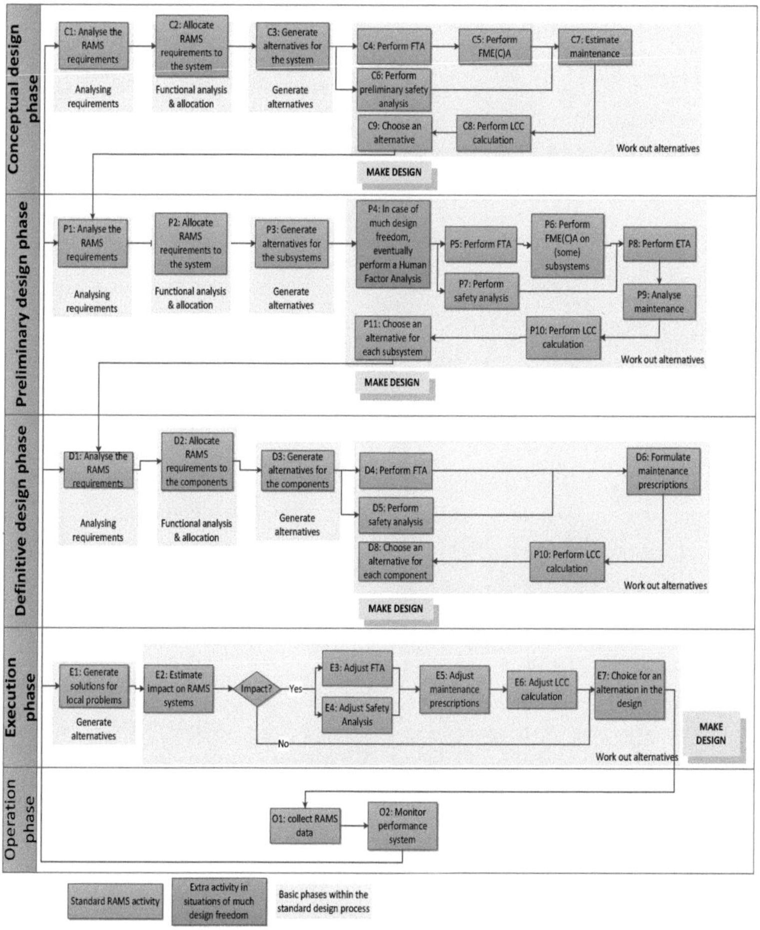

Fig. 162 RAMS procedure for contracts that includes maintenance (after [111])

K. Tzanakakis: *The Railway Track and Its Long Term Behaviour*, STTT 2, p. 297.
DOI: 10.1007/978-3-642-36051-0_38

39 A Case Study of Applications of RAMS and LCC in the Design Process

The project in the following example is based on a project described in [111] that has been modified to adapt –as an example- to railway projects.

Short description of the example project

The track of a railway line between City A and City B has to be constructed. The RAMS and LCC analysis will be presented for the design phase of the project, according to the model described in Chapter 38.

P1: Analysing requirements

The first step in the model for preliminary design phase is to analyse the RAMS requirements.

Within the project several requirements are being considered and these requirements can then be allocated to the track construction in this case.

P2: Functional analysis and allocation

In the second step the evaluation criteria that are used to evaluate several alternatives are described and the trade-off requirements can be determined.

The evaluation criteria within the project are:

- Reliability,
- Availability,
- Downtime,
- Initial cost,
- Maintenance cost and
- Cost for specific risks.

These evaluation criteria are derived from the RAMS requirements specified by the Infrastructure Manager. Within these criteria the engineering company develops the trade-off requirements that are used within the trade-off to evaluate the alternatives:

1. The reliability of the track should be 99,8%.
2. The availability of the track should be 99,0%.
3. The downtime should be less than 3600 hours.

For the contractor the lowest total costs are also important to gain a competitive advantage over other contractors. The total costs (in present worth) of the track can be divided into

K. Tzanakakis: *The Railway Track and Its Long Term Behaviour*, STTT 2, pp. 299–304.
DOI: 10.1007/978-3-642-36051-0_39

- initial construction costs,
- maintenance costs and
- costs for the risks that can occur during the life span of the track. A risk for example is that large maintenance needs to be done earlier than was expected in the design phase.

P3: Generate alternatives

Two different alignments have to be examined (Alignment A and B). Each Alignment alternative can be constructed as ballasted track, or as slab track. So the possible alternatives are:

Table 40 Alternatives in track alignment and track construction

Alternative 1	Alternative 2	Alternative 3	Alternative 4
Alignment A, ballasted track	Alignment A, slab track	Alignment B, ballasted track	Alignment B, slab track

P4: (optional) Human Factor Analysis

A human factor analysis is useful when human or human decisions influence the functioning of the system. In this case human factors have very little influence in the functioning of the track. Accidents on the track caused by human failure, for example, could influence the functioning of the track, but traffic accidents are excluded in the performance requirements. Also no data is available on how human factors influence the execution phase where some of the track characteristics are determined. Therefore, a Human Factor analysis is not performed in this case.

P5: Perform FTA

A Fault Tree Analysis (FTA) (see par. 34.7) can be done to model and analyse the failure processes of systems by graphically developing a tree starting from one top cause of a failure into smaller failure causes. In this case little knowledge exists about failure processes of track construction. Experienced designers of track have a global idea how and when track will fail in their life. They do not have specific data about the root causes of failures and therefore an FTA cannot be made in this case.

P6: Perform FMECA on subsystems

An expanded FMECA (par. 34.8) sheet is used to determine the effects of the main failure modes that can occur during the life of a track. Four main failure modes are identified using data from experienced designers in track construction. These four failure modes are:

1. Track geometry faults
2. Rail wear
3. Ballast wear
4. Subgrade failure

The occurrence of a certain failure mode is determined with the experiences from the past that have been identified by the track designers. The state of each failure mode will occur in practice in several stages which are identified in the Infrastructure Manager regulations.

The stages vary i.e. from A (high quality level) to F (Worst quality level) and in practice the Infrastructure Managers mostly demand a minimal quality level B of their track. Every state of the track below this quality is considered to be a failure and needs maintenance. When the quality level reaches below level B the effects of a failure mode can be determined by what is necessary to restore the required quality level.

With the occurrence and effect of a failure mode the risks for one failure mode and the total risk of all failure modes for one alternative can be calculated.

P7: Perform Safety Analysis

A safety analysis has to be performed according to the requirements of the Safety Management System of the Infrastructure Manager.

P8: Perform ETA

An Event Tree Analysis (ETA, see par.34.7) is not performed in this case due to the fact that events, such as accidents, that could cause the track to fail are excluded in the maintenance contract.

P9: Analyse maintenance

The maintenance period is 25 years, with, for track, a residual life of 5 years for the ballasted track and 25 years for slab track. The contractor is free to determine a maintenance strategy.

The maintenance plan for each alternative is partly based on the results of the FMECA that is done in P6. With the information of the effects of the failure modes for an alternative, the amount of downtime can be calculated and the number of failures during the maintenance period can be estimated.

The total operation time in this project is 262.800 hours[83]. Alternative 1 for example has an estimation of 57 failures during the maintenance period of 30 years. With this information the **reliability** of alternative 1 can be calculated as 99,98% (see par. 34.4.2). The downtime of alternative 1 follows from the estimated time that the maintenance for this alternative will take.

Availability is calculated with the uptime and downtime of a system (see par. 34.4.3). The expected downtime in this example for Alternative 1 is 2593 hours.

[83] 30 years X 365 days X 24 hours = 263.800 hours.

With this downtime and the total operation time, the uptime can be determined and the availability is calculated as 99,01%.

In this case downtime is transferred into cost with the use of the "lane rental principle"[84]. For this example, the contractor must rent busses for the passenger transfer in order to temporary close the line or change the timetable[85]. In this case the Infrastructure Manager demands a minimum amount of maintenance and disturbance for the railway user. Therefore when a line section is shut down for maintenance purposes a penalty of €3.000,- per hour is given to the contractor.

In Table 41 all alternatives are shown with the trade-off requirements that indicate the performance.

Table 41 Alternatives with performance trade-off requirements

Criteria	Alternative 1	Alternative 2	Alternative 3	Alternative 4
	Alignment A, ballasted track	Alignment A, slab track	Alignment B, ballasted track	Alignment B, slab track
Reliability	99,98%	99,99%	99,92%	99,99%
Availability	99,01%	99,67%	98,72%	99,57%
Downtime [hours]	2.593	864	3.371	1.123
Downtime cost	7.779.000 €	2.592.000 €	10.112.700 €	3.369.000 €

The other alternatives are calculated in the same way as the example above (Table 41).

P10: Perform Life Cycle Cost Calculation

The costs in this example are proportionally adjusted. In this step the relevant cost categories for this case are calculated, which are initial costs, maintenance costs and costs for certain risks.

[84] "Lane rental" is used by road projects to minimize the impacts of a project on the travelling public. It is a method of transferring the road user costs to the contractor. The contractor must rent a lane in order to close it. This creates a financial incentive for the contractor to minimize the duration of lane closures.

[85] An additional amount for compensation could be required due to the interruption of a seamless journey, and the travel inconvenience and travel time increase.

The initial costs and maintenance costs can be calculated with information from the contractor on how to construct the alternatives and the required maintenance. The risks are calculated on the basis of the results from the FMECA.

P11: Choose alternative for subsystem

On the basis of the calculated performance and total costs the four alternatives are evaluated. In Table 42, the criteria to evaluate the alternatives are shown.

Table 42 Evaluation of alternatives

Criteria	Alternative 1	Alternative 2	Alternative 3	Alternative 4
	Alignment A, ballasted track	Alignment A, slab track	Alignment B, ballasted track	Alignment B, slab track
Reliability	99,98%	99,99%	99,92%	99,99%
Availability	99,01%	99,67%	98,72%	99,57%
Downtime [hours]	2.593	864	3.371	1.123
Downtime cost [€]	7.779.000	2.592.000	10.112.700	3.369.000
Initial Cost [€]	20.000.000	25.000.000	15.000.000	18.750.000
Maintenance Cost [€]	2.593.000	864.000	3.370.900	1.123.000
Risks [€]	800.000	2.000.000	850.000	2.100.000
Total cost [€]	**22.593.000**	**25.864.000**	**18.370.900**	**19.873.000**

Table 42 shows that

- Alternative 3 is the overall cheapest.
- Alternative 4 is close to alternative 3, It is a bit more expensive than Alternative 3, but it has the best reliability and the second best availability and downtime characteristics.

For this specific case the best choice is considered to be **alternative 3**. It is the cheapest alternative and at the same time has a good overall performance.

40 LCC and WLC (Whole Life Costing)

40.1 What Is Whole Life Costing?

The terms Whole Life Cost (WLC) and Life Cycle Cost (LCC) have been used interchangeably – and their meanings have become confused.

WLC is a methodology for the systematic economic consideration of all whole life costs and benefits over a period of analysis, as defined in the agreed scope.

According to the draft of the ISO Standard 15686-5, WLC is equivalent to LCC plus external costs, thereby defining WLC as a broader term including within it life cycle costing and covering a wide range of analysis.

The difference between WLC and LCC is shown in Figure 163:

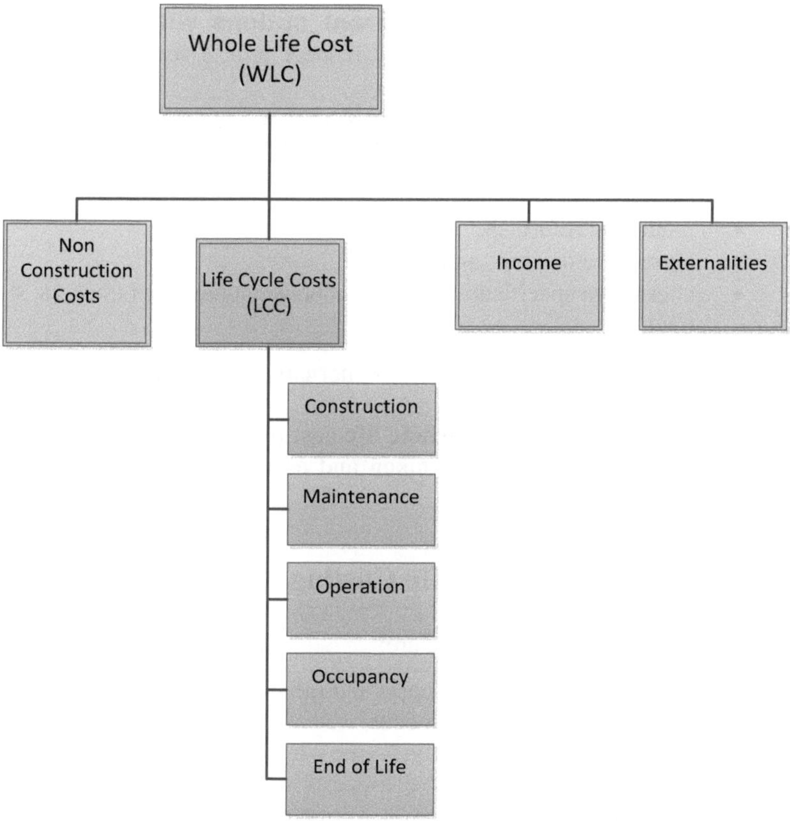

Fig. 163 LCC and WLC (Whole Life Costing) [86]

[86] BS/ISO 15686-5 Buildings & Constructed Assets.

K. Tzanakakis: *The Railway Track and Its Long Term Behaviour*, STTT 2, pp. 305–306.
DOI: 10.1007/978-3-642-36051-0_40

Whole Life Costing is defined in the draft International Standard, ISO15686 Part 5 as: *"economic assessment considering all agreed projected significant and relevant cost flows over a period of analysis expressed in monetary value. The projected costs are those needed to achieve defined levels of performance, including reliability, safety and availability"*.

Another definition easier to consider is[87]: *"the systematic consideration of all relevant costs and revenues associated with the ownership of an asset"*.

In the following, the definitions of LCC and WLC are given side by side in order to clarify the difference between both of them:

Definition of LCC (Chapter36)	*Definition of WLC*
the cost of an asset, or its parts throughout its life cycle, while fulfilling the performance requirements	*the systematic consideration of all relevant costs and revenues associated with the ownership of an asset*

Neither of these definitions reveals that **whole life costing is a tool, to assist in making decisions between different options with different cash flows over a period of time**. In this respect it is a form of **investment analysis**.

Whole life costing is relevant when considering whole estates, whole facilities, individual buildings or structures and when comparing alternative investment scenarios such as:

- retain and refurbish,
- alternative designs, and
- alternative specifications (such as between ballasted track vs. slab track).

It is particularly used to justify whether an alternative with a higher capital cost is justified.

The goal is to achieve an **initial whole life cost plan**, which can be modified as the design and construction is undertaken and as more details of specifications become available.

40.2 Basic Steps in Whole Life Costing

The basic steps in Whole Life Costing are:

1. Identify capital and operational costs and incomes
2. Identify when they are likely to occur
3. Use "discounted cash flow" analysis (see par. 36.3) to bring the costs back to a common basis.
4. Undertake sensitivity analysis of the variables such as the discount rate, the study period, the predicted design lives of components, assumptions about running costs, etc.

[87] See "National Platform for the Built Environment", UK.

41 Investment and Budgeting Decisions for High Track Quality

Track construction requires large investments.

The sole purpose of investing is a high initial quality. The sole purpose of maintenance is to assure that this initial quality leads to an extension of the service life of the track. Investment and maintenance should be viewed as elements of a strategy for the superstructure.

Lack of maintenance, delayed or improper maintenance of the track devalues the investment, as far as decreasing its service life[88].

41.1 The Structure of the Track Life Cycle Cost (LCC)

The track LCC can be analysed in

(1) Depreciation
(2) Cost due to operational hindrances
(3) Cost of track maintenance

Figure 164 gives an example of this cost structure. The mentioned cost elements are discussed below.

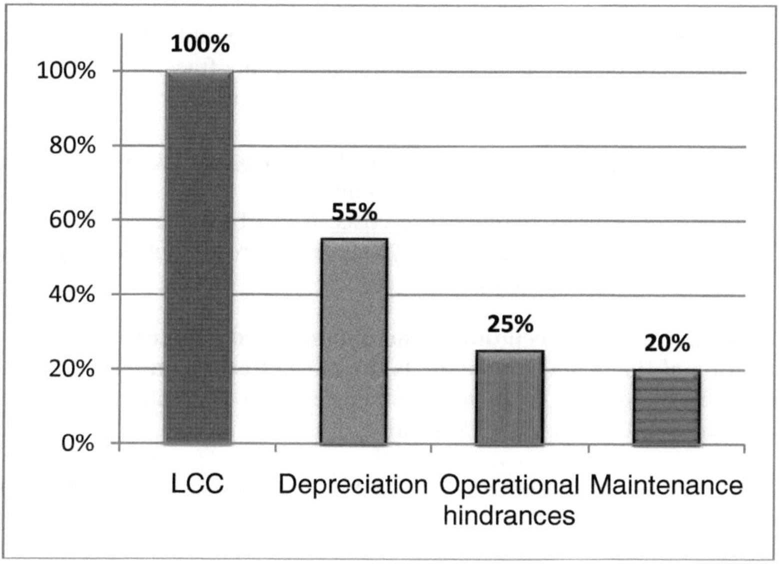

Fig. 164 LCC cost analysis (after [129] and after seminars of Prof. P. Veit[89])

[88] Prof. P. Veit, Technical University of Graz.
[89] Technical University of Graz.

K. Tzanakakis: *The Railway Track and Its Long Term Behaviour*, STTT 2, pp. 307–311.
DOI: 10.1007/978-3-642-36051-0_41

Investment costs - depreciation

An investment is an asset that a company owns and intends to use in operating activities. The railway track is an example. Generally the entire cost of acquiring an asset like the track cannot be deducted in the year acquiring it. Most assets lose their value over time (in other words, they depreciate), and must be replaced once the end of their useful life is reached.

Depreciating an investment means recovering the cost of the investment over a specified number of years, for the track, it is the service life (see par. 30.1):

$$\text{Depreciation} = \frac{\text{LCC}}{\text{Service Life of the track}}$$

Cost of operational hindrances

Cost of operational hindrances arises from disturbances of the traffic operation i.e. due to unplanned maintenance, which costs to the Infrastructure Manager (less revenues from access charges, penalties to the railway undertaking etc.).

Maintenance costs

The maintenance costs include planned and unplanned maintenance.

41.2 Strategies to Decrease the LCC

In order to reduce the LCC of the track following measures can be taken:

1. Decrease of the depreciation cost. It is possible by decreasing the numerator or increasing the denominator of the fraction (Equation in par. 41.1), that means by decreasing the LCC of the track or by increasing its service life.
2. Decrease of the cost of operational hindrances. It can be achieved by better planning of the planned maintenance.
3. Decrease of track maintenance costs. It can be achieved by applying an optimized maintenance strategy (methods, proper outsourcing, optimum choice of the track components etc.).

Due to the fact that depreciation is the dominant cost factor, a strategy for a prolongation of the track service life is the most effective one (Figure 164).

We will examine some cases where strategies with cost cuttings in short-term, lead to an increase of the track LCC (long-term).

Case 1: Decrease in depreciation – Increase of the track service life, accepting its degradation

This strategy results in the imposition of speed restrictions that result to an increase of the costs of operational hindrances. LCC increases: lengthening the useful life of the line against quality is uneconomical (Figure 165).

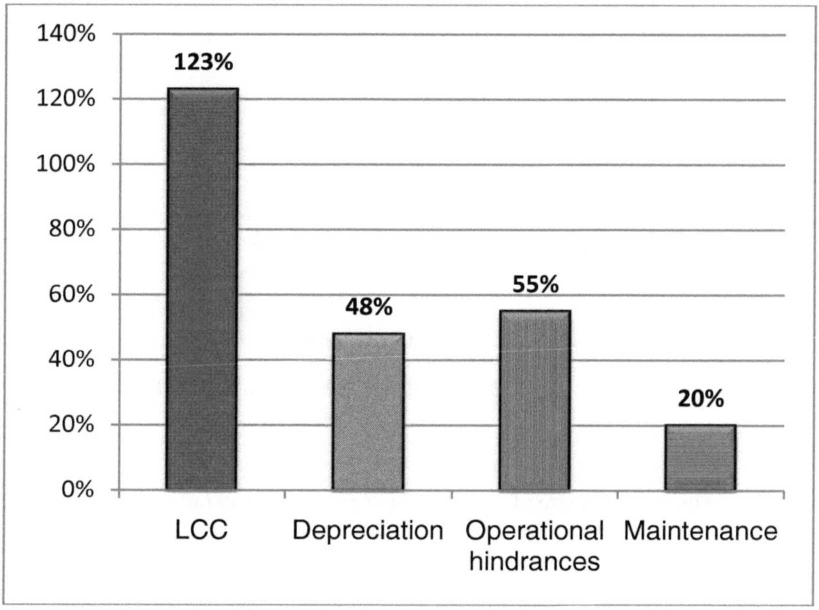

Fig. 165 Extension of service life due to speed restrictions (after [129])

Case 2: Decrease of the track maintenance volume

By reducing the required maintenance[90], the track service life is reduced, thus increasing the cost of depreciation (the denominator of the fraction in the equation in par. 41.1 is decreased), see Figure 166.

---

[90] i.e. by reducing the available budget for maintenance.

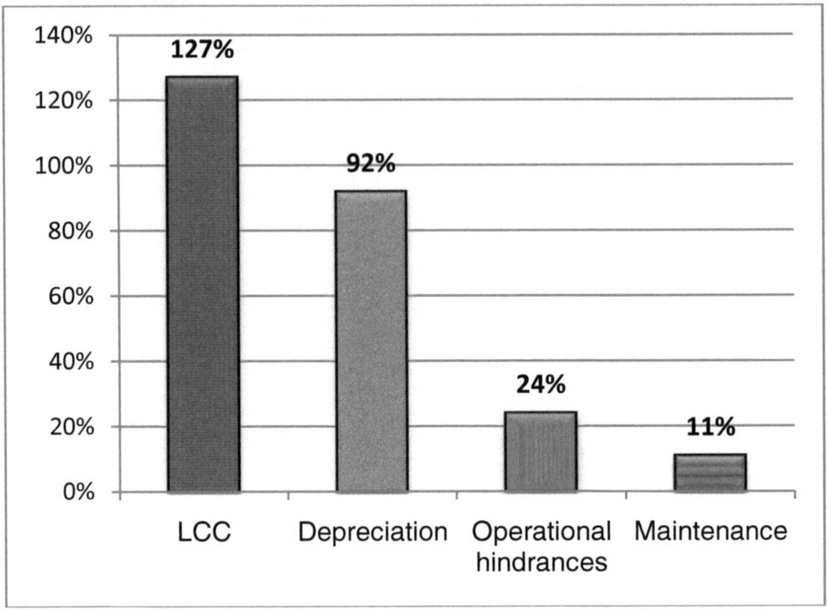

Fig. 166 Decrease of the track maintenance budget (after [129])

Conclusion

Figure 167 gives an overview of the various maintenance strategies and their impact on the track LCC.

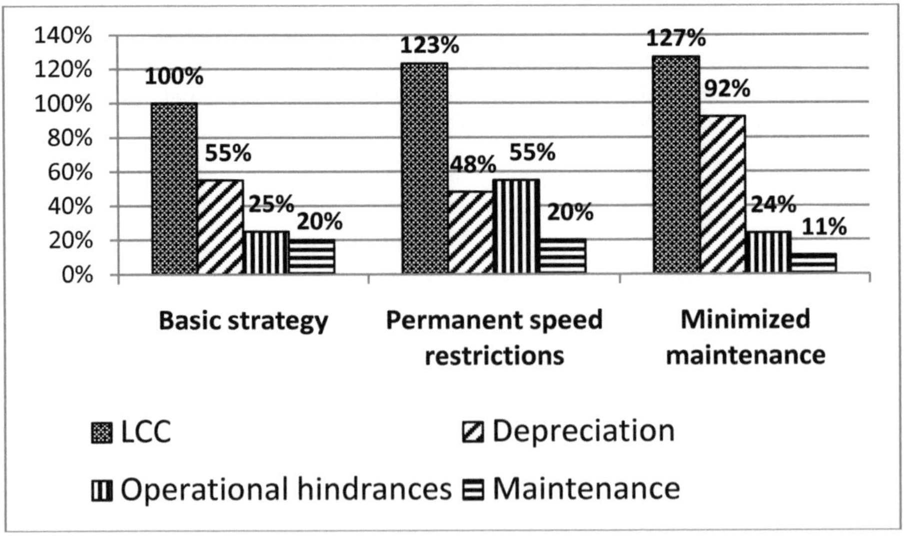

Fig. 167 Overview of the various maintenance strategies (after [129])

The main conclusions drawn from what is presented here are:

- **A track with good quality is the most economical one.**
- **But only a proper maintenance can ensure the benefits of a track with an initial good riding quality.**

Section VIII: Outsourcing Track Maintenance

(IN-HOUSE VERSUS OUTSOURCING)

Do what you do best and outsource the rest.

Outsourcing of maintenance services is not a definite path to maintenance excellence. Through outsourcing it could be possible to cut cost and make the maintenance more effective.

In this section, advantages and disadvantages of outsourcing, factors influencing decision making, types and characteristics of outsourcing contracts, and cases of applied outsourcing are topics discussed among others.

42 Introduction

> *Key terms:*
> *Strategic and tactical outsourcing, contractor's performance, types of contracts, Performance-based Contracting (PBC), Fixed-cost-incentive-fee & Cost-plus contracts, alliance relation contracts, risks associated with contracting, outsourcing scenarios, modelling costs*

Traditionally, governments own public infrastructure such as railway, roads, communication networks, public buildings, and other public facilities. Usually maintenance of these infrastructures is done in-house. However, there is a growing trend towards outsourcing these maintenance activities to external agents.

Outsourcing is a service contract of maintenance actions where components or system failures are rectified by an external contractor (agent) for an agreed period of time. The contractor charges a price for such service. His profit is influenced by many factors, such as the reliability of the components or system, servicing strategies to be taken during the period, and the policies of the contract [119].

Outsourcing of maintenance isn't an automatic strategy to achieving maintenance objectives. The outsourcing of maintenance services is not a definite path to maintenance excellence. Through outsourcing it could be possible to cut cost and make the maintenance more effective.

In the following, advantages and disadvantages of outsourcing, factors influencing decision making, types and characteristics of outsourcing contracts, and cases of applied outsourcing are topics discussed among others.

K. Tzanakakis: *The Railway Track and Its Long Term Behaviour*, STTT 2, p. 315.
DOI: 10.1007/978-3-642-36051-0_42

43 From Tactical to Strategic Outsourcing

Outsourcing has been engaged by industry as a useful tool to improve their efficiency. However, the main reason why organizations employed outsourcing was in the 1980's less strategic, and more tactical.

According to a survey (1991), the biggest benefit that organizations expected by introducing outsourcing was a reduction of overhead and short-term costs [123]. Such a simple task as the cleaning of office buildings, distributing internal mail and so forth had been outsourced.

In the middle of the 90's, tactical outsourcing moved in some areas toward **"strategic outsourcing"** and the organizations obtained considerable benefits. Strategic outsourcing is completely different from conventional outsourcing, because it **is aligned with the organization's long-term strategies**. Strategic outsourcing is not just a simple and useful tool, but in fact a complex one.

Management should carefully consider ;

- their future core competencies,
- organization structure,
- cost structure,
- performance level, and
- competitive advantages,

before making any decisions about "strategic outsourcing" (see Table 43)

Table 43 Comparison between Tactical and Strategic Outsourcing

	Tactical Outsourcing	Strategic Outsourcing
Main purpose	Problem-solving	Restructuring organization
Duration	Short term	Long term
Decision rights	Not transferred	Transferred to providers
Impact on organization	Small	Large

K. Tzanakakis: *The Railway Track and Its Long Term Behaviour*, STTT 2, p. 317.
DOI: 10.1007/978-3-642-36051-0_43

44 Cost and Benefit of Outsourcing

Outsourcing can bring a certain amount of benefit to the organization. Outsourcing also has advantages and disadvantages. Thus, careful consideration is required before making decisions about introducing outsourcing.

In some countries (United Kingdom, Netherlands) routine maintenance is fully contracted out. In the Netherlands, for example, this maintenance is contracted out to contractors by so-called "output process contracts". In these contracts a precise description is given of what is wanted with respect to quality. Contractors are responsible to achieve this quality. The Dutch rail Infrastructure Manager does not tell the contractors how to carry out the maintenance, but evaluates the quality by several modern measuring instruments, and by monitoring and analysing disturbances. This means that in these countries the contractors are responsible for the diagnosis of routine maintenance [77].

44.1 Advantages

Maintenance cost saving

Cost saving is the first benefit. Of course maintenance costs will never drop if the productivity of the external agent is exactly the same as that of the Infrastructure Manager (IM). It is a prerequisite that a contractor has a higher efficiency and productivity than the IM that decides to execute outsourcing.

In practice, the contractor will apply cutting edge technology, state-of-the-art equipment, experience, and well-trained highly motivated personnel. Moreover, if the economy of scale works in service provider's activities is applied, maintenance cost will decrease even more.

Enhancing organization's core competencies

Through outsourcing, and making use of cost savings, Infrastructure Managers can focus on their core competencies and allocate their resources in order to enhance their core competencies, which is the desired outcome of a successful outsourcing strategy.

Improve performance and quality

If the contractors have high skills, Infrastructure Managers can expect not only cost reduction, but improvement of performance and service quality. If an Infrastructure Manager has only limited experience and skills, outsourcing can compensate for their weakness by utilizing the expertise of outside providers.

K. Tzanakakis: *The Railway Track and Its Long Term Behaviour*, STTT 2, pp. 319–323.
DOI: 10.1007/978-3-642-36051-0_44

Risk sharing

Contractors are responsible for their work, and they own risks related to their activities. Transferred risk is proportional with the span of the outsourced area. This is a significant advantage for the Infrastructure Manager, especially in cases with a relatively high risk.

44.2 Disadvantages

Outsourcing can bring benefits, as presented. Outsourcing is quite a useful tool, however, once misused, or abused, it could cause irreversible fatal damage to the sustainability of an Infrastructure Manager, as described below.

Loss of competencies

There is a danger that the competencies and capabilities of the Infrastructure Manager are being transferred unexpectedly to the contractors. The main reason of this tragedy is the misunderstanding of their core competencies. In particular, the Infrastructure Manager should be careful on this point, because, fundamentally, only non-core competencies area should be outsourced.

The Infrastructure Manager may also lose control over maintenance. It is easier to check that certain maintenance works have been carried out, than to check whether certain quality levels are achieved. In the United Kingdom this disadvantage has led to considerable problems. In the Netherlands, modern measure instruments are used among others to check the quality so as to overcome this disadvantage.

The rail Infrastructure Manager may also lose a feeling for the costs. Since routine maintenance is contracted out, it is unclear for the rail Infrastructure Manager what exactly the costs for the contractors are, to realize the contracted quality levels. This makes the negotiations more difficult [120].

Loss of cross-functional synergies

Cross-functional synergies of activity sectors of the Infrastructure Manager have to be checked before the final decision of outsourcing (non-core activity sectors may have cross-functional synergies with core competencies).

Transition cost

Outsourcing involves a lot of administrative work in the transition phase, if resources such as employees, facilities, and equipment are transferred to contractors. It may result in considerable costs that must be considered.

If the power of the labour union is strong and influential it's not a simple task due to the fact that organizations have to spend a certain degree of effort to negotiate with them.

Uncertainty

It is very important to assure a good performance level of the contractors. Inferior contractors can destroy the perception of the Railway Undertakings (customers) for the Infrastructure Manager, from which it is so difficult to recover.

Difficulties in reversing

Once outsourced, it is very difficult to reverse and operate in-house. The Infrastructure Manager (organization) should think that reversing is the last countermeasure even if the outsourcing has failed.

44.3 Make or Buy? Decision Making about Outsourcing Track Maintenance

The advantages and disadvantages of outsourcing have been presented. The big question is how to decide about "buy or make" and the choice of their sections to be outsourced?

For the theoretical discussion about outsourcing, four sets of factors to make decisions are introduced and discussed [120];

(1) strategic factors,
(2) market factors,
(3) service and technology factors, and
(4) economic factors

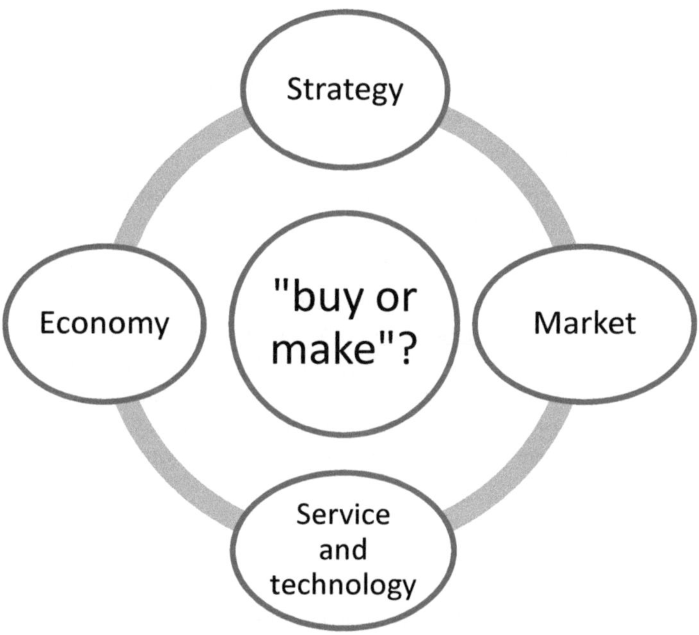

Fig. 168 Factors influencing decisions of "buy or make"

44.3.1 Strategic Factors

The Infrastructure Manager has to identify its core competencies before deciding on an outsourcing strategy in each railway organization.

The failure in identifying the core competencies would be critical to the establishment of a successful outsourcing strategy. Again, core competencies should be kept in-house; otherwise, the Infrastructure Manager will lose its competitiveness in the future. Non-core business should be chosen as work for outsourcing.

It should be noted that the identification of core competencies depends a lot on each railway's CEO's philosophy.

44.3.2 Market Factors

In each country there are limited number of contractors, so in general, there isn't a hard competition between the railway maintenance contractors. Due to this fact it might not be always easy to find well-qualified maintenance contractors. The existence of quality of maintenance contractors is an important factor.

If the contractor isn't well-qualified, railway should maintain internally; otherwise, outsourcing can be chosen.

If healthy competition exists, then it might be advisable to outsource (provided that –as mentioned- there are well-qualified maintenance contractors).

44.3.3 Service and Technology Factors

There is a number of maintenance activities, which enables **partial outsourcing**. Partial outsourcing is much easier for the Infrastructure Manager, especially for the first step of launching an outsourcing strategy, because contracting structure can be limited. For instance,

a. measurement of the track geometry (by a track recording vehicle)
b. execution of repair work
c. inspection and repair for bridges

could be outsourced to enhance the efficiency of the track.

In this sense, track maintenance can be broke down into pieces, enabling partial outsourcing, which might be easy for the railway to apply. If the service is highly modular (as a big part of the track maintenance is), it enables the Infrastructure Manager to more simply outsource portions of the service. On the contrary, if the service is highly integrated, it is much more difficult to outsource portions of the service.

44.3.4 Economic Factors

Cost reduction is the most hard-pressed problem in railway maintenance.

Labour costs can be expected to be reduced by introducing outsourcing, because, in general, railway employees are protected by the labour union, which disables flexible management in terms of labour. Once a job is outsourced, the contractor should be able to manage the employees more cost effectively than those in the railway.

Infrastructure managers should be careful in terms of cost estimation and make decisions only when total cost could be reduced. Even if direct labour costs were reduced through outsourcing, the other costs incurred should not be neglected. If these factors are all taken into consideration, maintenance can be outsourced.

Summarizing, cost reduction is one of the most significant benefits that the Infrastructure Manager is supposed to enjoy by introducing outsourcing; thus this point should be carefully verified before making a final decision. Of course, the outsourcing plan should be approved only when the total estimated cost of outsourcing is lower than that of owning it [120].

45 Measuring Contractor's Performance

After signing the contract with an appropriate external agent (contractor), the next concern is to monitor performance, since a certain amount of responsibility is transferred to the contractor.

The relationship between the individual performance and the metrics is described in[121].

> *"The link is simple. If a firm measures a, b, and c, but not x, y, and z, then managers begins to pay more attention to a, b, and c. Soon those managers who do well on a, b, and c are promoted or are given more responsibilities. Increased pay and bonuses follow. Recognizing these rewards, managers start asking their employees to make decisions and take actions that improve the metrics. Soon the entire organization is focused on ways to improve the metrics. The firm gains core strengths in producing a, b, and c. The firm becomes what it measures...."*

This description warns those Infrastructure Managers trying to introduce performance metrics to evaluate outsourcer's performance. If the settings of performance metrics are not appropriate, outsourcers will be misled by them and result in unexpected performance that will be different from what Infrastructure Manager expected at the beginning of contract.

In particular, simple metrics are preferred in terms of implementation; however, a single performance metric is usually not enough to manage outsourcers, because the scope of outsourced work is not so simple that it can be measured by using a single scale. Multiple measurement scales are preferred in most cases.

Therefore, generally speaking, special attention is required to effectively choose combined performance metrics, which cover whole activities that outsourcers are required to do.

Infrastructure Managerexpects high quality of service with lower costs. In order to analyse the performance accompanied with cost effectiveness, it's necessary to illuminate these attributes about maintenance activities: inputs, outputs, outcomes, and value for the Infrastructure Manager as a customer (Figure 169).

K. Tzanakakis: *The Railway Track and Its Long Term Behaviour*, STTT 2, pp. 325–327.
DOI: 10.1007/978-3-642-36051-0_45

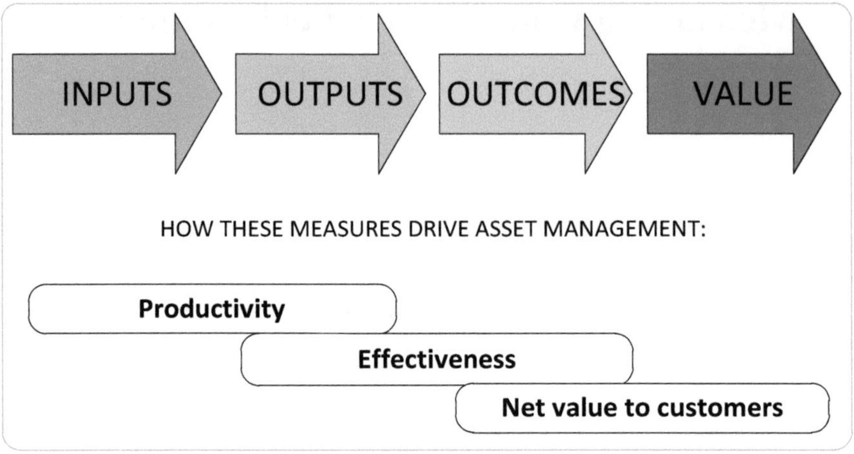

Fig. 169 Attributes about maintenance activities (after [120])

45.1 Inputs

Inputs: all the resources that contribute to the production and delivery of outputs. Inputs are **"what we use to do the work"**.

In railway maintenance, the inputs consist of labour, equipment, and materials. The funds needed to pay for these resources may also be considered as an input [120].

45.2 Outputs

Outputs: the final products, or goods and services produced for delivery. Outputs may be defined as **"what we produce or deliver"**.

In track maintenance, examples of output measures are length of track maintained, the number of switches maintained, or the restoration of rail breaks [120].

45.3 Outcomes

Outcomes are the effects and changes that are a result of the outputs. Outcomes may be defined as **"what we wish to achieve?"**. For example, an outcome of track maintenance could be a better track quality [120].

The ratio of outcomes relative to inputs can be calculated to address the effectiveness of a program. Typical measures might be an outcome per labour hour, per equipment hour, or per Euro of expenditure (see Chapter 56).

One might also examine cost effectiveness (see par. 36.2), which is the cost per unit of outcome achieved. As one transitions from using output measures to outcome measures, the emphasis shifts from productivity to effectiveness.

45.4 Values (or Impacts)

Values (or impacts) are customer-oriented outcome measures expressed in terms of the value received by the customer (results).

46 Types of Contracts and Its Characteristics

46.1 Introduction

(based on [120])
According to a dictionary, the word "contract" is *a legal document that states and explains a formal agreement between two different people or groups, or the agreement itself.* Each prospective party, who is going to make a contract, has an expectation toward another party, and negotiates with each other on monetary and environmental conditions, and finally, if it is acceptable, reaches a formal agreement. A contract is a product of these rigorous processes[120].

Firms (here the Infrastructure Managers) need to rely, by and large, on the others' resources and skills to achieve their business goals; thus contracting is a necessary skill in order to survive. One thing we should mention is that there are various kinds of contracts to be chosen in terms of

- pricing methods (pricing methods vary from fixed-price to cost- plus)
- relationship types (relationship types range from traditional to alliance)
- ordering types (ordering methods consist of method-based and performance-based).

Each method has strengths and weaknesses respectively.

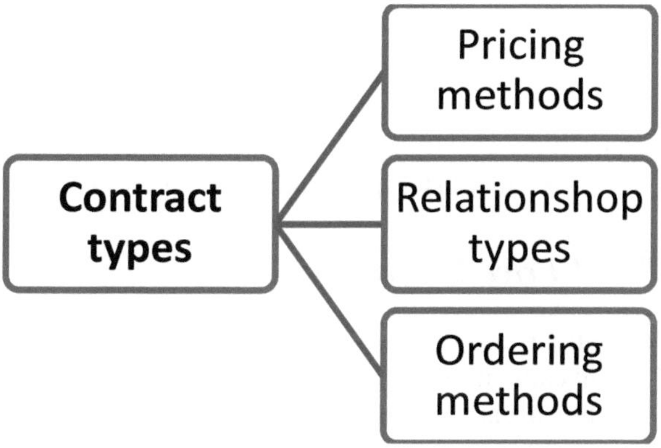

Fig. 170 Contract types

Choosing the appropriate kind of contract that is tailored to the situation is a key maximizing the benefit from a contract. In this chapter, we will discuss the fundamental issue about contracting, characteristics of various types of contracts, and the appropriate combinations to enhance the benefit of contracting.

K. Tzanakakis: *The Railway Track and Its Long Term Behaviour*, STTT 2, pp. 329–336.
DOI: 10.1007/978-3-642-36051-0_46

46.2 Contracts According to Types of Pricing Method

Contract types vary according to

- the degree and timing of the **responsibility** assumed by the contractor for the costs of performance; and
- the amount and nature of the **profit incentive** offered to the contractor for achieving or exceeding specified standards or goals.

The specific contract types range from **fixed-price,** in which the contractor has full responsibility for the performance costs and resulting profit (or loss), to **cost-plus-fixed-fee,** in which the contractor has minimal responsibility for the performance costs and the negotiated fee is fixed. In between are the various **incentive contracts,** in which the contractor's responsibility for the performance costs and the profit or fee incentives offered are tailored to the uncertainties involved in contract performance (Figure 171)[120].

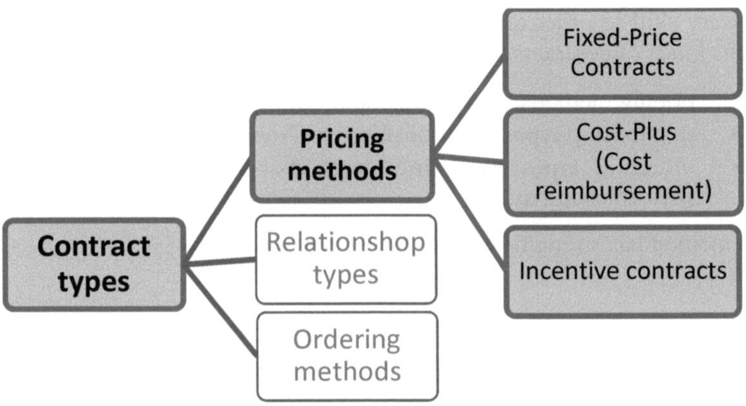

Fig. 171 Types of pricing methods contracts

46.2.1 *Fixed-Price Contracts*

Fixed-price contracts are contracts without any adjustment during the term of the contract. Careful estimation of the total cost in advance is very important, however overpayments are usually required, that usually causes problems related to this payment method. Adjustments are usually based on established prices, actual costs of labour or material, or cost indexes of labour or material.

Fixed-price contracts are generally appropriate for services that can be defined objectively and for which the risk of performance is manageable. Simple service provisions, are suitable for fixed-price contracts.Therefore, employing this method for a complex service that involves several high risks is not suitable[120].

46.2.2 Cost-Plus (Cost-Reimbursement) Contracts

A cost-plus (cost reimbursement) contract enables contractors to be reimbursed for all the cost that is incurred during the contract period. A cost-plus contract is suitable for contracts that involve high risks for contractors, such as research and development (R&D) for totally new technology, or a material supply contract under highly fluctuating raw prices. When uncertainty is too high, contractors cannot expect enough profit under a fixed-price contract, and then contractors will not even try to join the bidding. Cost-reimbursement contracts are preferred in such situations[120].

46.2.3 Incentive Contracts

Incentive contracts are appropriate when a firm-fixed-price contract is not appropriate and the required services can be acquired at lower costs and, in certain instances, with improved delivery or technical performance, by relating the amount of profit or fee payable under the contract to the contractors' performance: the final payment is not exactly fixed in advance, but will be determined in accordance with the quality of performance that the contractor has implemented during the contract period. If the quality of performance doesn't meet the expectations, some degree of penalty should be forced. Negative incentives are also required to motivate the contractors.

There are two major tools to activate the incentive scheme, **cost incentives** and **performance incentives:**

- First, if the actual cost for the contractor is below the estimated cost, the saved cost can be shared with the Infrastructure Manager and the contractor based on an adjustment formula. A contractor, thus, will be encouraged to perform in a cost effective manner. Most incentive contracts include only cost incentives.
- Second, performance incentives, even though they are not utilized often, may be considered in connection with specific elements of the contractor's performance. Positive and negative performance incentives shall be considered in connection with service contracts for performance of objectively measurable tasks, when quality of performance is critical and incentives are likely to motivate the contractors.In addition, when the Infrastructure Manager's significant concern is the time of delivery, delivery incentives can be considered.

When the Infrastructure Manager tries to structure a multiple-incentive arrangement, it should motivate the contractor to try for outstanding results in all incentive areas. However, incentives can conflict with each other. For example, when very attractive performance incentives that overcome cost incentives are provided, the contractor can focus on performance improvement neglecting some cost overruns. If the Infrastructure Manager's primary objective was low cost, the result would not be in accordance with the Infrastructure Manager's expectation. Therefore, arranging multiple-incentives to balance each component is essential to the success of contracting[120].

46.3 Pricing Types and Its Characteristics

The use of fixed-price contracts has been traditionally, limited to projects with relatively few technological and economic uncertainties. On the other hand, cost-plus contracts avoid the problem of overpayment by Infrastructure Managers, but the Infrastructure Managers subjects itself to the problem of potentially higher costs, because in this scheme there is a limited incentive for the contractors for cost reduction.

To control the negative incentives of cost-plus contracts, it has become common practice in the US to replace standard cost-plus contracts with "cost-plus-fixed-fee contracts", so that the firm's profit allowance is fixed rather than being proportional to actual project costs. "Cost-plus-incentive-fee contracts" are also used nowadays, in which the contractor's profit increases proportionally with cost.

As a summary, the relationship between these pricing methods and risk allocation is shown in Figure 172.

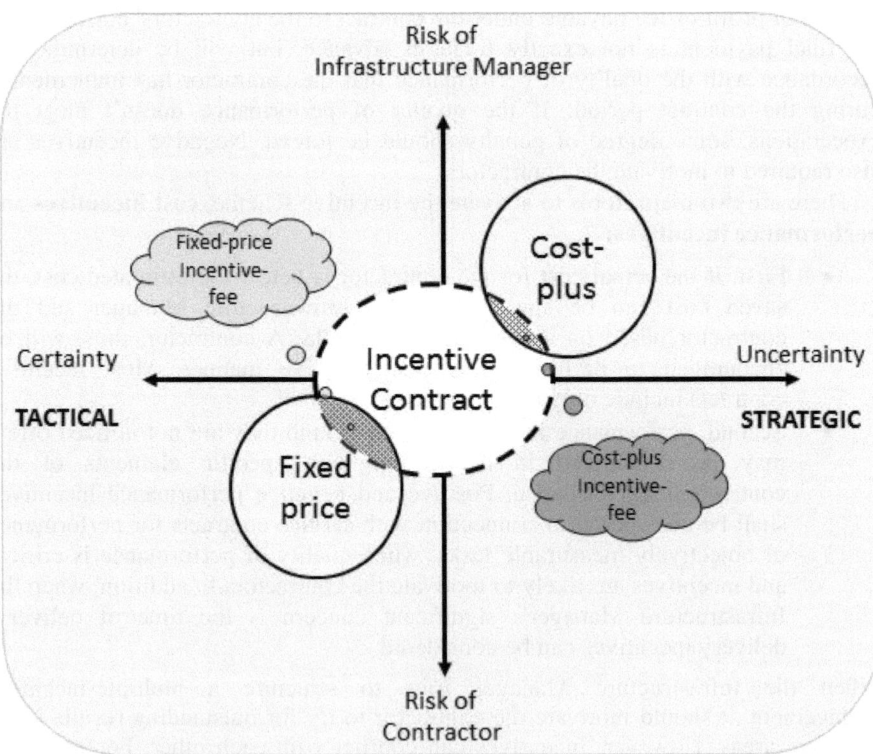

Fig. 172 Relationship between three major pricing methods and risks (after[120])

Projects with a certain amount of risk and uncertainty (due to the lack of experience like totally new research and development) are likely to be priced under cost-plus contract.

Relatively simple tactical contracts(without a risk to be taken care) are likely to be priced under fixed-price contract.

The problem is how both parties reach agreement in terms of the understanding of involved risk. If it is agreed and settled in an inappropriate form of pricing method, both parties will never feel comfortable during the contract term[120].

46.4 Contracts According to Relationship Types between Infrastructure Manager and Contractors

Takinginto accountthe relationship between the Infrastructure Manager and the contractors, contracts can be categorized as ([120])

(1) traditional,
(2) cooperative,
(3) partnering, and
(4) alliance contracting.

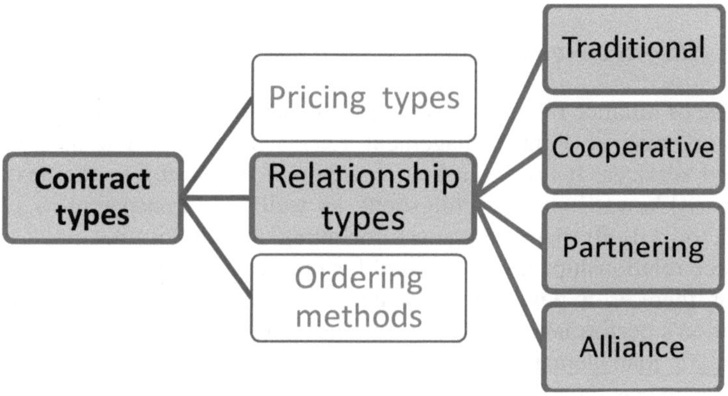

Fig. 173 Types of contracts according to relationship types

46.4.1 Traditional Relationship

The legal and contractual framework between Infrastructure Managers and contractors creates a strong compliance/control relationship relying on extensive checking and verifying of the contract. This type of contract is appropriate where the potential for variation during the contract period is low, and the risk of failure is also low. Disadvantages of this type are poor communication, lack of continuity between projects, time-consuming administration in processing variations, claims and negotiations and a tendency for a confrontational relationship.

46.4.2 Cooperative Relationship

Cooperative relationships bring the contracting parties closer together and involve both parties in cooperative management of the contract. This approach incorporates principles, such as developing trust, obtaining commitment and improving communication, but does not extend to some of the higher principles such as risk sharing. This style of relationship can be used as a step to find out whether there is potential to develop the relationship further.

46.4.3 Partnering Relationship

Partnering is a formal process supported by both a legal contract and moral commitment by key stakeholders and other parties to act in the best interests of each other. The basis of partnering is: <u>together we can solve problems and maximize opportunities</u>. This relationship has existed in the construction industry in Australia for over 10 years[120].

Partnering relationship can promote contract effectiveness and efficiencies through: commitment by both parties based on common objectives, equity and trust, implementation of joint strategies for developing mutual goals and continuous and joint evaluation.

46.4.4 Alliance Relationship

In the case of alliance relationships the Infrastructure Manager and the contractors accept a degree of joint management responsibility for the delivery of the prescribed services. It takes the key elements of partnering to the next level by having a risk/reward sharing philosophy as well as a transparent or open-book approach towards all financial matters, including cost and profit.

Alliance relationships are best suited for providing services that are <u>difficult to define</u> or there is a possibilityto change significantlyover time, critical to an organization's performance or requiring <u>innovative solutions from the contractors and creative management by the Infrastructure Manager</u>. Long-term, strategic contracts when each party relies on each other are suitable, too.

As a tool for measuring performance, key performance indices (KPI) (see par. 56.3) are usually defined in advance according to the characteristics of service. These indices are also used as evaluation for the contractor's contribution, which will result in bonuses and penalties for contractors.Alliance is employed to achieve the following three objectives[122];

- To provide extra leverage for the organization's core competencies in order to deliver long- term sustainable competitive advantage;
- To move the organization into long-term and developing commitment to new markets, territories, or technologies that were previously closed to it; and

- To provide a platform for the kinds of organizational learning that is central to its future business success, but would otherwise be unobtainable.

Figure 174 maps the appropriateness of these types of relations according to the risks and likelihood and significance of failure in the contracted service.

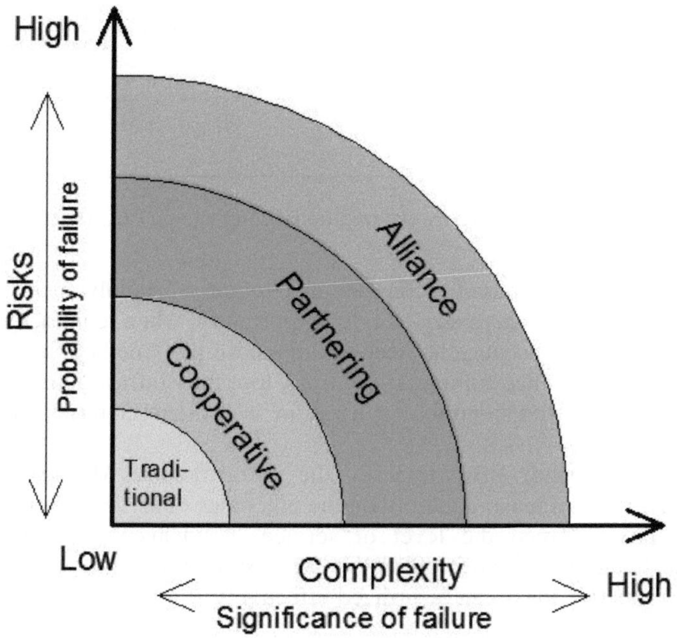

Fig. 174 Relationship Styles, Magnitudes of Risks, and Complexity of Service [resource; ANAO, 2001]

Risks and complexity associated with the contracted work are the key variables to determine the appropriate contract structure.

46.5 Ordering Types: Method-Based and Performance-Based Contracting

In some areas today, the performance-based contract (PBC) is becoming common.PBC is a contract that defines performance expectations in terms of outcomes or results (see also 47.1). In contrast, the traditional method-based contract defines methods and processes to produce the outcomes. The focal point shifts from output to outcome (par. 45), as shown in Figure 175.

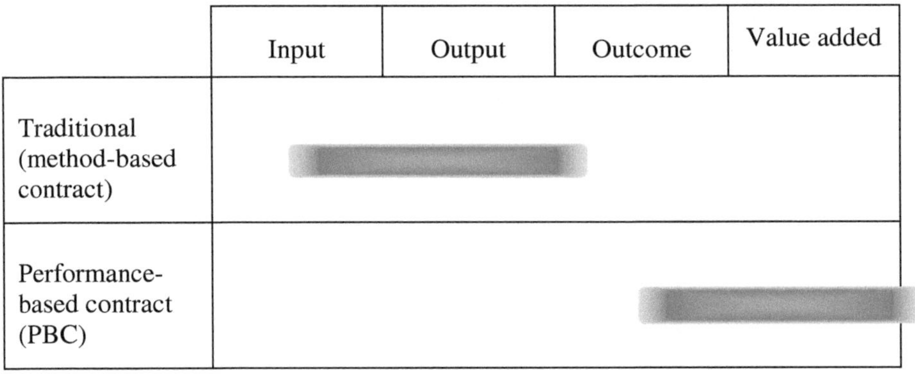

Fig. 175 Different focus between traditional contract and PBC (after [120])

Traditional method-based contracting requires involvements of the Infrastructure Manager in every single opportunity when contractor's service provision takes place, because he needs to diagnose the "method" of the service. Usually, quality assurance follows every time, too; thus Infrastructure Manager is involved frequently and contractor's freedom is limited in terms of choice of method.

On the other hand, PBC releases the Infrastructure Manager from this administrative work, focusing mainly on the outcomes of service. Prerequisites are that both parties agreed the level of service in advance and only qualified contractors should be chosen. Contractors are free to choose the working method; thus their productive ideas can be utilized effectively.

47 What Kind of Contracting Is Suitable for Track Maintenance?

Let's make the hypothesis that the Infrastructure Manager decides to outsource its track maintenance to an external contractor. The next concern is what kind of contracting he should choose to ensure that the outsourcing is successful.

47.1 Performance-Based Contracting (PBC)

It is the management practice of transferring or subcontracting some functions, activities and responsibilities performed in-house to external agents with a specified level of service to be met [123].

A performance based maintenance contracting is an approach to contracting that provides incentives and penalties to the contractor to achieve specified targets for measurable outcomes and outputs [125]. The performance measures are related to the condition of the different asset type and also to the outcomes of maintenance on operation, safety and economy.

There are different levels and different scenarios of maintenance outsourcing. The three key responsibilities differentiating these different scenarios are who determines:

(i) What asset type to be maintained
(ii) When to maintain
(iii) How to maintain.

The different scenarios are presented in Chapter 49.

The decision of outsourcing could also take into account the responsibilities of the Infrastructure Manager and the contractor in the maintenance process. When contractors have major responsibilities to decide what, when and how questions, is referred to as the performance based maintenance. A remarkable feature of performance based maintenance is that the Infrastructure Manager does not specify the maintenance technique but rather defines the performance requirement of the maintenance function. It is often called functional guarantee contract, performance based contract, output based contract, full service maintenance, performance specified maintenance contract, total maintenance contract, performance contract etc.

A **performance-based contract** is better than a traditional method-based one in terms of simplicity and usefulness. In strategic outsourcing, outsourcers are not only providing specific services written in a contract, but they have a wider range of rights about their work. In other words, they have rights to choose the way of performance by making their own decisions.

K. Tzanakakis: *The Railway Track and Its Long Term Behaviour*, STTT 2, pp. 337–341.
DOI: 10.1007/978-3-642-36051-0_47

In order to measure the outsourcer's performance, it is quite important to set effective packages of metrics that measure the performance.

PBC should be applied only when service provider (contractor) possesses enough competitiveness.

The adoption of performance based maintenance contract isn't widely accepted in the railway maintenance but is a well-established practice in road maintenance. A successful deployment of performance based maintenance contracting in the railway system will require a robust performance management process: a critical system in any performance management process is a performance measurement system.

47.1.1 Motivation for Performance Based Maintenance Contracting

The motivation and advantages of performance based maintenance contracting as identified by national cooperative highway research program [125] include (see also par. 44.1):

1. Potential to reduce maintenance costs
2. Potential to improve level of service
3. The transfer of risk to the contractor
4. Incentive for practical innovation
5. Better competence and skills with higher availability
6. Enhanced asset management
7. Change in performance criteria from a focus on inputs and outputs to customer-oriented outcomes
8. Reduction in contract administration requirements
9. Promise of stable financial obligation and reduced lifecycle cost.

47.1.2 Issues of Performance Based Maintenance Contracting

The fundamental issues to be addressed in outsourcing of maintenance activities were discussed in [124][91]. In railway infrastructure maintenance, these issues could be technical, management, economic or legal.

It is very important for the Infrastructure Manager to carry out a proper evaluation of the implications of outsourcing their maintenance. If done properly, outsourcing can be more effective than in-house maintenance, facilitating the achievements of maintenance objectives and greater business productivity.

Track maintenance possesses a certain number of possibilities in applying performance-based contracting. The reasons are;

- Outcomes of track maintenance are more important than methods, and
- There are several measurable indices for outcomes of track maintenance.

[91] The readiness of the company to outsource, what to outsource and six practical steps to evaluate the outsourcing process are the major issues addressed.

A traditional method-based contract is the most reliable and steady way for transportation infrastructure maintenance, because each action, such as inspection and repair, is double-checked by the railway and by contractors. However, it requires a lot of contribution for Infrastructure Managers in contract execution.

Incentive Scheme in PBC

As noted before, a planned and controllable part of maintenance should be priced under fixed-price- incentive-fee contract whose formula can be suggested as follows:

Payment = Fixed price + Base Profit + Incentive Profit Incentive Profits

= a * f_1 (Infrastructure quality) + b * f_2 (Operation delay caused by infrastructure) + c * f_3 (Cost incentive)

Where, a, b, c are variables to weigh these factors.

As an incentive part, not only cost incentive, but quality of infrastructure and contribution to service stability should be considered. The incentive formula can be customized based on each railway's philosophy; however, incentive should be directly connected with the desired outcomes that the railway expects, rather than outputs or other indirect factors.

Figure 176 shows the examples of the formula to figure out the incentive fees, rewards and penalties, in each component. In setting each formula, it is important to pay attention that negative incentives should not be so large that contractor can lose money. The primary aim of an incentive scheme is to increase the contractor's performance, rather than punishing contractors' inferior performance.

Of course, there could be another set of incentive schemes. However, the railway should be careful in choosing the combination of components linked with incentives, in order to build comprehensive and moderate incentives which are also accepted and agreed upon by the contractors.

In addition, hardship factors should be taken into account for incentives, especially when multiple contractors are involved.

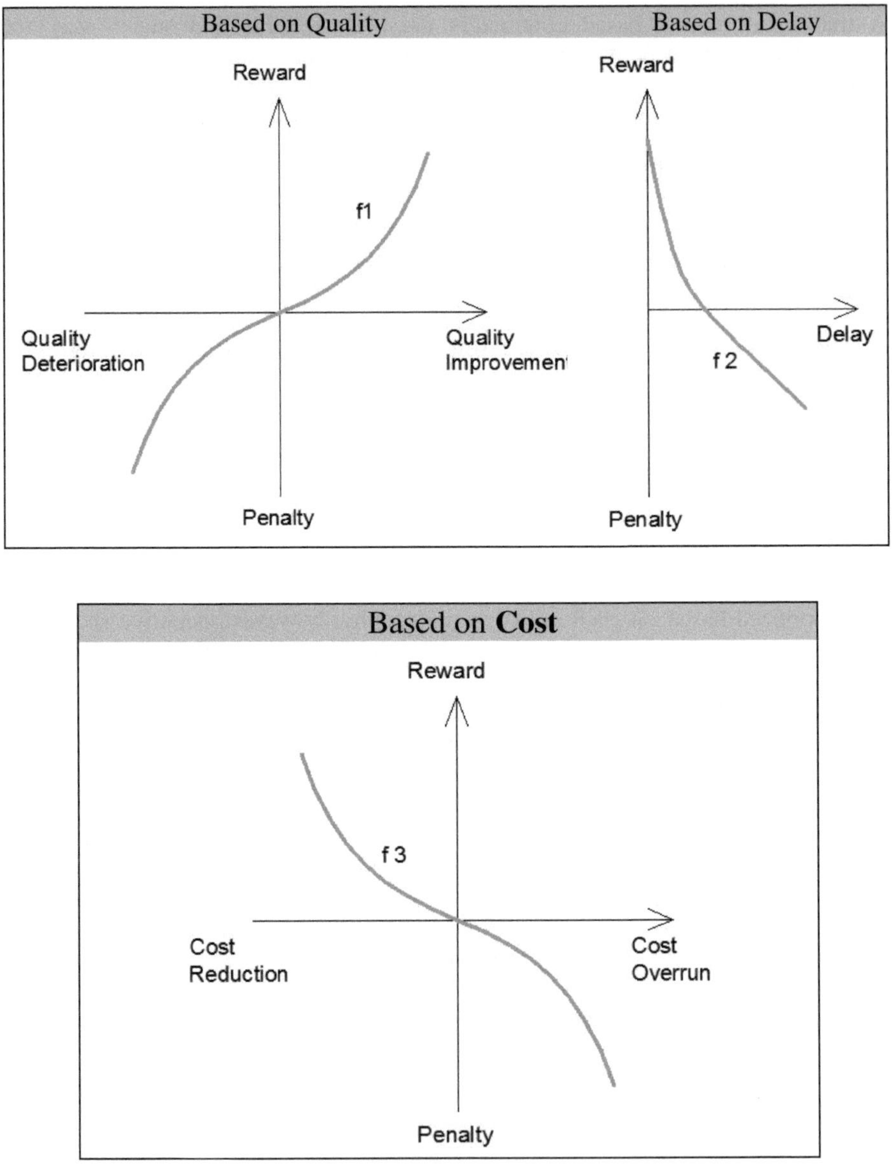

Fig. 176 Example of incentive formulas (conceptual) (after [120])

47.2 Contracts with Combination of Fixed-Cost-Incentive-Fee and Cost-Plus

If the railway is going to outsource most of maintenance, the combination of fixed-cost-incentive- fee and cost-plus would be suitable in terms of pricing. The reason is the wide variety of track maintenance, ranging from planned inspections and repair to unpredictable incidents. In general, the former can be priced under fixed-cost basis, because its uncertainty is lower than the latter, which should be priced cost-plus basis due to its high uncertainty. Moreover, there are several measurable indices in evaluating outcomes of track maintenance, which will be mentioned later (par. 56).

47.3 Contracts Based on Alliance Relation

In terms of relationship, an alliance relation is better, because

- Long-term perspective is essential in track maintenance, and
- Both parties need information accumulated individually by each party.

First, track maintenance requires a long-term perspective, thus a single year contract is not suitable. Long term contracts (5 year or 10 year) are appropriate, if a reliable contractor is found. Long-term contracts would encourage contractors to hire high qualified staff and invest in high-quality machinery and equipment which may be amortized in a contract period.

Second, the maintenance contractor's information about the current condition of infrastructure is crucial to make decisions about capital investment for upgrading infrastructure.

In conclusion, performance-based contracting with contractors whose relation is "alliance contract" under the combination of fixed-price and cost-plus basis fits the railway track maintenance outsourcing.

48 Common Risks Associated with Contracting

Well-managed contracts can bring significant benefits to an organization. Risk management is among the important issues in contract management. The following table presents some steps for risk management[120].

1st step: **Completely understood of the contract**	The first step in risk management associated with contract management is ensuring that the contract is completely understood by all parties. The three key issues are: • the output that the contracted service supports; • the critical success factors to the delivery of the outputs • the internal input necessary for the delivery of the outputs.
2nd step: **Risk assessment**	Risk assessment is the second step. There are two kinds of risks that should be clarified, contract risk and contract management risk. **Contract risk** is the risk associated with the delivery of the service, in other words, the probability that the service will not be delivered in accordance with the requirements of the contract in terms of time, cost, quality and quantity. Most of these risks arise externally, which may be considered "uncontrollable". On the other hand, **contract management risk** arises from within the organization. In general, this type of risk is less likely to occur; however, it may lead to the erosion of the contractual relationship and ultimately adversely affect service delivery.
3rd step: **Risk analysis**	After the identification of risks, the third step is risk analysis. The magnitude of the impact given by each negative event should be figured out in order to understand the vulnerability of the system in terms of service provision. Some alternatives can be suggested to reduce the impact and enhance stability of service by accurate risk analysis.
4th step: **Risk treatment**	The final step is risk treatment. The Infrastructure Manager may accept, control, reduce, or transfer these risks identified and analysed.

K. Tzanakakis: *The Railway Track and Its Long Term Behaviour*, STTT 2, p. 343.
DOI: 10.1007/978-3-642-36051-0_48

49 Maintenance Outsourcing Scenarios

The parties involved in the development of the outsourcing scenarios are, as already discussed:

- Owner of equipment: Infrastructure Manager (IM)
- Service agent: Contractor

We can distinguish three main decision categories for outsourcing depending on the responsibility given to the service agent:

- ► D1: **What** (components) need to be maintained?
- ► D2: **When** should the maintenance be carried out?
- ► D3: **How** should the maintenance be carried out

Three basic scenarios can be formulated:

Scenarios	Decisions	
	Infrastructure Manager	Contractor
S1	D1, D2	D3
S2	D1	D2, D3
S3	-	D1, D2, D3

Scenario S1 — The IM decides when and what should be maintained. He has his maintenance strategy (i.e. see par. 32.11), he inspects the track and dictates the maintenance activities. The contractor has to decide about the maintenance method and the organization of the work (incl. needed resources).

Scenario S2 — The IM decides what should be maintained. He has his maintenance strategy (i.e. see par. 32.11), he or/and the contractor inspects the track and the maintenance activities have to be determined. The contractor has to decide about the time to execute maintenance, choose the appropriate method and organize the work (incl. needed resources).

Scenario S3 — The IM sets the maintenance standards. The contractor has full responsibility for the track maintenance.

K. Tzanakakis: *The Railway Track and Its Long Term Behaviour*, STTT 2, p. 345.
DOI: 10.1007/978-3-642-36051-0_49

50 Modelling Costs of Service Contract

Total costs of service contract may include the cost of planned preventive maintenance and corrective maintenance in the form of minimal repairs and failure replacement, cost of inspections and condition monitoring, cost of risks, and penalties for failure to meet agreed safety, reliability and availability of requirements[119].

Therefore, the total expected costs C of service contractis

$$C = C_s + C_i + C_d + C_r + C_p$$

Where,

Cs= total expected costs of maintenance services
Ci = expected costs of inspection
Cd = costs of downtime
Cr = cost of risks associated
Cp = penalties for failure to meet contract agreements.

K. Tzanakakis: *The Railway Track and Its Long Term Behaviour*, STTT 2, p. 347.
DOI: 10.1007/978-3-642-36051-0_50

51 Infrastructure Data Management in Case of Outsourcing of M&R Activities

Data management when outsourcing certain important activities is crucial. If an Infrastructure Manager (IM) has contracted out the maintenance and renewal work to one or more contractors, the task of the Infrastructure Manager concerning the collection of infrastructure data will change. Apart from some visual spot checks of the track from time to time, the IM does not itself collect the infrastructure data needed. It has to rely on the data collected and presented by the contractors. The role of the IM turns into one of managing the data collection, instead of performing it itself.

Therefore in the case that maintenance and renewal (M&R) is outsourced, the Infrastructure Managershould ask, apart from the above-mentioned aspects, the following additional questions:

- Who is obtaining the data?
- How is the data collected?
- Who is storing the data?
- Is the stored data centralised or decentralised?
- Who is analysing the data?
- What is the quality of the data?
- Who is checking the quality of the data?
- How often will the information be updated? (How often should it be updated?)
- Are there mixed interests with the contractors concerning the availability (and the place of storage) of infrastructure data?

K. Tzanakakis: *The Railway Track and Its Long Term Behaviour,* STTT 2, p. 349.
DOI: 10.1007/978-3-642-36051-0_51

52 Cases of Outsourcing Applied

52.1 Case 1: Performance Based Railway Infrastructure Maintenance in Sweden[92]

The experience in Swedish Transport Administration "Trafikverket" could be traced back to 1988 when the Swedish Railways was vertically separated into two parts: the Swedish National Rail Administration, with responsibility for the infrastructure management; and the Swedish Railways with responsibility of running train services.

The responsibility of the Swedish National Rail Administration involves services and maintenance works on the infrastructure to keep the asset in a reliable, available and safe state. At this time the maintenance and renewal works were done with in-house resources with little or no engagement of private or external agents in the engineering works.

The restructuring process proceeded in 1998 when the infrastructure administration was horizontally separated into two parts: the infrastructure administration named Banverket, and Banverket Production which had the task of maintenance, rebuilding and new construction of the railway infrastructure. This structure was often referred to as client-contractor, which involves a service buyer - infrastructure administration, and a service provider – maintenance contractor. Since 2010, the maintenance service provider has become an independent state company competing with other private contractors in the railway infrastructure maintenance contract market.

The contracting out of maintenance works to competitive market started in 2001 with few players and few contracts. The type of outsourcing differs in volume of work and responsibility of the contractors. The involvement of the maintenance service providers (the contractors) in the maintenance process also vary. However at the early stage of the outsourcing, the labour and equipment of the contractors were utilized while most of the engineering analysis and technical function were done in house by the infrastructure manager. The evolution of the maintenance function in recent years has seen a greater involvement of contractors in the maintenance management of the infrastructure. Some infrastructure regions within Swedish Transport Administration now employ outcome based or functional contracting strategy. Figure 177 shows a picture of the evolution of maintenance organization with different scenario in timeline and parties involved.

---

[92] See [123].

K. Tzanakakis: *The Railway Track and Its Long Term Behaviour*, STTT 2, pp. 351–354.
DOI: 10.1007/978-3-642-36051-0_52

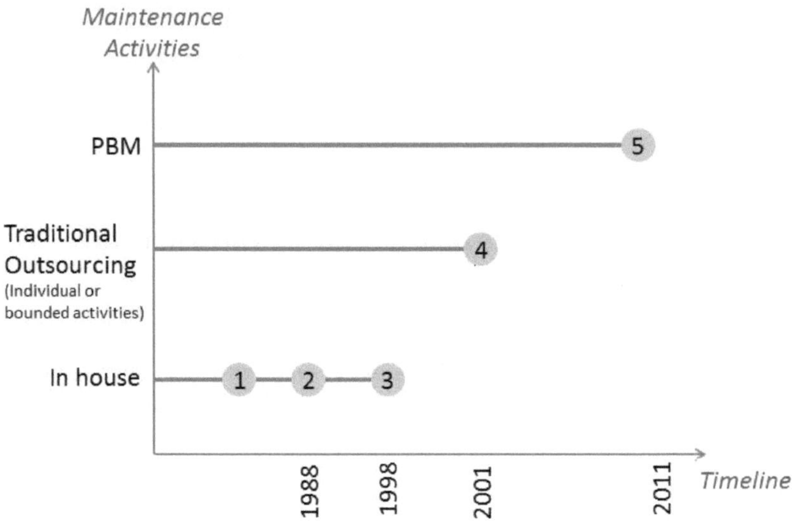

Fig. 177 Evolution of railway infrastructure maintenance at Swedish Transport Administration [123]

PBM = Performance Based Maintenance

❶ *All operations done by SJ*[93],
❷ *Traffic operations by SJ and infrastructure management by Trafikverket,*
❸ *Maintenance outsourced to a public company,*
❹ *Competitive contractor market with single or bundled activities being outsourced,*
❺ *Full maintenance service outsourced with performance specification.*

In this new trend in maintenance outsourcing, the contractors become largely involved in the maintenance process not only at the operational level but also right in the strategic level.

52.2 Case 2: The British Experience – From Railtrack to NetworkRail [94]

The development of maintenance as an outsourced activity grew rapidly in the 1990s, with Railtrack being one of the major organisations to adopt this policy in full. Railtrack outsourced tens of thousands of directly employed maintenance personnel to a number of contractors – predominantly civil engineering companies operating in the construction sector. Railtrack's new approach to maintaining its infrastructure was widely criticised, the more so after a number of high profile

[93] The largest train operator in Sweden.
[94] See [126].

maintenance-related accidents, in particular those at Hatfield in October 2000 and at Potters Bar in May 2002. The government then decided to establish Network Rail, a "not-for-profit" mutual company, as a replacement for Railtrack. One of the early decisions Network Rail made was to take back the maintenance of the London-to-Reading section in-house.

Within a few months, Network Rail decided to insource all maintenance across the whole of the rail network, and to re-employ the personnel associated with it.

In recent years, the rail industry has not been alone in driving an outsourcing policy for maintenance then reversing it – there are also examples in the electricity sector. The water industry has been more cautious in its drive to outsourcing and the results have been mixed. Benefits have been achieved on the "below ground" assets.

For "above ground" assets, and in particular mechanical and electrical maintenance, there have been few examples of demonstrably effective contractual arrangements.

Railtrack's task was to maintain the infrastructure and invest in its development with the Strategic Rail Authority (SRA). How, then, could one of its early decisions be to outsource the direct maintenance work – its core task?

The answer is that this was an act of self-deception: a straight-forward cost cutting measure to deliver immediate benefits in operational expenditure (OPEX)[95] with no real understanding of the impact on capital maintenance expenditure (CAPEX). And that was not the only impact on railway performance. As well as the thousands of former British Rail employees who were outsourced to contracting companies, many experienced and committed employees were lost to the industry altogether through redundancy and early retirement. These were the people who knew how things worked.

Mike Levery, the author of the article "From Railtrack to Network Rail" [126] refers: "*In March 1998, I was asked to assist in the resolution of an issue in a partnering arrangement between Railtrack and two infrastructure contractors. Railtrack had established a partnering arrangement with the two contractors, outsourced the employees and established new contractual arrangements. Railtrack then informed both companies that the volume of work was to be reduced by 40%, and only one company would be required in future. My role was to achieve an amicable way forward between all parties within the spirit and objectives of the arrangements that had been established.*

Despite the inauspicious circumstances, an agreement was achieved. However, what caused me the greatest concern was Railtrack's decision, with no obvious rationale in terms of the impact on asset performance, to cut the volume of work by such a large amount. There was bound to be an immediate financial benefit, but what were the long-term implications of a decision such as this? How would the recently outsourced workforce feel towards Railtrack, having been transferred to contractors and then faced with redundancy through reduced work volumes? It is easy to deliver short-term savings on maintaining infrastructure assets that are

[95] An **operating expenditure** or **OPEX** is an on-going cost for running a product, business, or system. Its counterpart, a capital expenditure (CAPEX), is the cost of developing or providing non-consumable parts for the product or system.

inherently robust, but the long-term effects will always surface, sometimes sooner than expected.

I spoke at a conference in December 1998 on the subject of "win/win partnerships" at a forum of potential rail maintenance contractors .... I was asked in the Question and Answer session how contracting companies should respond to Railtrack's request to develop the asset management strategy for the network. I must admit to having been astounded by this question. No matter what asset management model you look at, the development of the asset management strategy is a long way away from the day-to-day maintenance work carried out in the field. My response was to question whether this requirement was a contractual obligation and, if not, it could not be considered the responsibility of the contractor. It is right that maintenance contractors should support the development of their client's asset management strategy, but not to develop it for them. Alarm bells were ringing! The demise of Railtrack was not unexpected, particularly as the government had lost confidence in a Plc[96] *which continued to demand significant increases in public funding whilst delivering a perceived poorer service.*

Further, the media presented Railtrack's contractual arrangements as being poorly managed, with contractors' personnel inadequately trained for carrying out maintenance work on the railways. It was only a matter of time before an alternative to Railtrack would be sought.

The decision to establish Network Rail was, in my view, the only option for the rail infrastructure, as it removed the drive to reduce OPEX[97] *to meet shareholder aspirations and therefore cut costs on maintenance. Network Rail's focus is purely to maintain the infrastructure efficiently without needing to seek cost-cutting initiatives to deliver shareholder value. .... Network Rail now had the freedom to raise its capital resources to meet the railway's long-term infrastructure investment needs."*

[96] Public limited company.

[97] An acronym for operational expenses.

53 Lessons Learned – Advices for Outsourcing

In the following, some lessons learned from ProRail (Netherlands) experience and advice for outsourcing are given [127]:

1. **You have to know your business and be in control of it**:
   i. Only if you can specify and measure it, can you manage it.
   ii. Skilled people and dedicated instruments are needed to manage costs and performance.
   iii. Know your infrastructure, costs realised and minimum quality accepted.
   iv. Reorganise your processes, instruments and skills to control improved results.

2. **Do not outsource maintenance if you don't know how to stay in control of costs and performance.**
3. **Outsource step by step: plan, separate, split-off and outsource.**
4. **Outsourcing should not be a goal but a way to improve.**
5. **Outsourcing needs a cultural change and takes time**.

Keep in mind:

The market will not give you anything for free.
Outsourcing is not a solution for your financial or managerial problems.
Tasks are outsourced, but the end-responsibility for the overall result not.

K. Tzanakakis: *The Railway Track and Its Long Term Behaviour*, STTT 2, p. 355.
DOI: 10.1007/978-3-642-36051-0_53

Section IX: Performance and Cost Effectiveness of the Railway Track

The infrastructure as the main production factor in the railway system's value chain accounts for a significant part of the full system costs. Therefore, it is very important for Infrastructure Managers to have a good control of infrastructure management in general and of costs in particular to improve competitiveness.

Aspects that are discussed in this section include guiding principles for an effective railway track, methods to measure track performance, benchmarking as a practical tool for improving performance by learning from best practices and understanding the procedures by which they are achieved. Furthermore, a method for estimating costs related to maintenance and renewal of the track are also addressed.

Section IX. Performance and Cost Effectiveness of the Railway Track

54 Cost Factors of Track Construction and Maintenance

Key terms:
Cost factors (construction and maintenance), performance measuring, Key Performance Indicators (KPI's), benchmarking, Infrastructure data management, maintenance cost modelling

The infrastructure as the main production factor in the railway system's value chain accounts for a significant part of the full system costs. Therefore it is very important for Infrastructure Managers to have good control of infrastructure management in general and of costs in particular to improve competitiveness.

The following table gives the factors that shape the cost of the superstructure and explains their significance.

Table 44 The cost shapers of railway superstructure [93]

Cost factors	Importance
Initial quality of the superstructure	The initial quality determines the behaviour of the track superstructure throughout its service life. The initial quality can never be restored through maintenance.
Quality of superstructure	The life cycle cost (LCC) of the superstructure is affected by different quality grades of the superstructure. In lines with low traffic, and poor superstructure, the life cycle cost increases 2 to 3 times compared to lines with good superstructure, and especially lines with a heavy traffic load, the increase is up to 8 times higher.
Density of switches	A switch with a radius of the divergent track with R = 500 m and a length of about 50 m, corresponds to the life cycle costs of a track with a length of 450 m. Therefore the least possible switches have to be installed in the track.
Curve radii	The life-cycle costs of a curve radius R = 250 m, is about three times compared to a corresponding segment of tangent track.
Cost of operational hindrances	The cost of operational hindrances due to operation in a lower performance level as its actual potential or due to execution of maintenance works can reach a value of 30% of its life cycle cost (LCC) for lines with a heavy traffic load.

K. Tzanakakis: *The Railway Track and Its Long Term Behaviour*, STTT 2, pp. 359–360.
DOI: 10.1007/978-3-642-36051-0_54

Table 44 (*continued*)

Traffic load	Traffic load affects significantly the cost. The effects are not important as long as superstructure is correctly sized. In this case the cost increases proportionately. As long as the superstructure does not correspond to the traffic passing over it, the cost increases over-proportionally.
Quality of rolling stock	The life cycle cost (LCC) is affected in approximately 5% if locomotives are friendly to the track and reaches 10% in extreme cases.

55 Guiding Principles for an Effective Track

> ***Efficiency is doing things right; effectiveness is doing the right things***
> Peter Drucker

Guiding principles for track construction, maintenance and renewal activities to assure an effective railway track are given below([14] and [129]):

1. Try for an initial high track quality

The most important cost driver is the initial quality. For permanent way this means especially a high quality demand of subgrade. Subgrade quality dominates the behaviour of track. Therefore an excellent substructure quality is a precondition for any optimization.

a. Subsoil quality

According to information provided by SBB (Swiss), ÖBB (Austria) and Technical University of Graz/Austria, subsoil rehabilitation in connection with a well-functioning drainage has the biggest effect to track quality (up to 800%). The standard should be new construction with a long service life and a high initial quality. Initial construction costs are about 50 % of rail-borne construction methods and the service life can be doubled from 40 to 80 years. PSS (formation protective layers) is always a compromise solution and should be chosen only where asphalt is not possible.

b. Ballast

Quality, standard and material of ballast has a big influence on Maintenance & Renewal activities and is a key factor to determine the moment of renewal. Ballast is the critical element limiting service life of track. Optimization within the superstructure requires good ballast quality. This analysis indicates that sustainable track designs require lower stresses in the ballast bed [129].

c. CWR (continuous welded rails)

A track with continuous welded rails has a longer service life and reduced maintenance, but needs a good track quality (ballast profile, geometry), therefore Infrastructure Managers prefer continuously welded rail in all scenarios (except, for instance, in areas prone to mining subsidence and too narrow curves).

2. Keep water away from the track surface

Crucial principle is the drainage of the track body and substructure to guarantee stability. Never invest in substructure without draining the water. Remove water from the side of the track as soon as possible using a suitable water drainage system. Even the best subgrade is badly affected as soon as the

K. Tzanakakis: *The Railway Track and Its Long Term Behaviour*, STTT 2, pp. 361–362.
DOI: 10.1007/978-3-642-36051-0_55

drainage system is poor and water is kept in track. Drainage is an integrated part of the technical structure “track”.

3. **Make an effective design and as simple as possible**

a. Use standardized track componentsfor a homogenous track.
b. Use track components with low maintenance costs or maintenance free and with a long service life.
c. Use appropriate materials according to local conditions for achieving optimal results (speed, load, radius, line gradient, etc.)
d. The number of switches (turnouts) is a main cost driver. Total cost of switches (turnouts) reach values 9 to 13 times higher than open track, counted in cost per length of through-going track.
e. Within the alignment the radius is the cost driver, while the gradient seems to have less importance. Especially radii less than 300 m increase track cost three times compared to straight sections.

4. **Maintain when required**

a. Maintain to the correct standard.
b. Use best maintenance practices.
c. Replace technologically obsolete or non-productive track components.
d. Remove track faults/damage at an early stage to avoid accelerating the wearof the trackfrom damage (i.e. through preventive grindingetc.).

5. **Be environmentally astute**

Take into account any environmental impacts caused by track work. In particular pay attention to poor maintenance practices and inadequate drainage which could cause sediment and erosion issues.

6. **Train staff**

Ensure staff is well trained and kept up to date with current best practice.

7. **Respect and keep historic values**

Where the track has distinctive historic values, ensure that this character is maintained.

56 Measuring the Performance

> *"Measurement is the first step that leads to control and eventually to improvement. If you can't measure something, you can't understand it. If you can't understand it, you can't control it. If you can't control it, you can't improve it"*
>
> H. James Harrington
>
> *"You get what you measure. Measure the wrong thing and you get the wrong behaviours"*
>
> John H. Lingle
>
> *"You cannot manage what you cannot measure"*

56.1 Measurement of Performance

"You can't manage what you don't measure". It is an old management adage that is accurate today. Unless you measure something you don't know if it is getting better or worse. You can't manage for improvement if you don't measure to see what is getting better and what isn't.

Measures of performance have been used by management for a long time to review current operational capabilities. Such measures have been used to assess both departmental and corporate performance, as well as trend performance achieved against established goals and objectives.

It is a good practice to regularly provide management with accurate information regarding key performance indicators. Such information should be measurable and used to assess performance and identify areas requiring management attention. Overall indicators relevant to performance of organization, indicators to measure progress in achieving goals and objectives, and specific indicators for monitoring current performance problems and performance in specific functional areas are the common practice in many industries. Information is presented in a way that provides recognition of trends and comparison of actual versus expected results and, where appropriate, clearly indicates corrective action and the results of these actions.

56.2 Performance Indicators

An indicator can be defined as a number that describes the performance for a specific activity or occurrence [132]. The number is normally related to a fixed measurement scale, e.g. the thermometer showing -17° C or the speedometer's 75 km/h.

K. Tzanakakis: *The Railway Track and Its Long Term Behaviour*, STTT 2, pp. 363–366.
DOI: 10.1007/978-3-642-36051-0_56

If the indicators are extended to measure the performance of working life related issues, they are called performance indicators; they can be classified in to seven horizontal main groups, i.e. efficiency, effectiveness, productivity, budget/profit, quality of working life, innovation, and quality ([132]), see Figure 178. The indicators can later on be aggregated to a performance index.

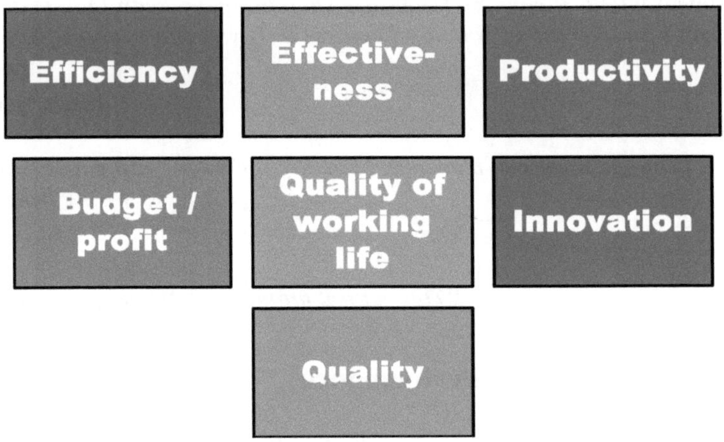

Fig. 178 Classification of Performance Indicators[132]

The use of key variables and performance indicators is about management and control of physical assets and processes on different levels inside an organization. They provide alarm signals when something in the outcome measures differs from the expected value: it is important not only to get early warnings when things are not going in the desired direction but also to be able to diagnose why it is occurring.

56.3 The Importance of Key Performance Indicators (KPI's)

Key Performance Indicators(KPI's) are "key" because they measure what is critical to the success of the company. In many cases managers are not even clear what constitutes success. This applies throughout industry and commerce, but it is particularly common amongst companies which have been for many years dependent on government subsidy, such as railway infrastructure management companies (IMs)[77].

Success consisted merely in satisfying government and doing whatever they wanted the company to do – which might change with a change of government. It is difficult even to have a vision or strategy, when it is so dependent upon factors outside management control. In leading commercial companies on the other hand, top management ensures that the vision drives the company, and that as many employees as possible understand exactly what the vision is and what constitutes success.They are aware what the different measures of success are and what is critical to creating that success. For them, it is easy then to select key indicators.

The purposes of KPIs include the following:

- To identify priority areas for improvement within each IM
- To identify the best performing IMs in the different success dimensions: These will be the ones to emulate.
- To help determine how much improvement is possible
- To monitor progress over time: if performance has improved, what practices have led to the improvement?
- To monitor an IM's own progress over time, in order to determine if it is progressing towards its vision and objectives.
- To estimate the effect of differences in processes and practices on actual outcomes
- To set targets for improvement

56.4 A Case: Banverket Maintenance Performance Indicators

Today, there exist a number of different measures and asset condition parameters in Banverket (Sweden) that are used or planned to be in use for follow up purpose regarding maintenance activities.

In Banverket handbook BVH 824.10 following indicators are described:

- Rail head top wear - % of track distance
- Rail head side wear - % of track distance
- Rifles and waves - % of track distance
- Rail defects – number of defects per track kilometre and year
- Rail defects - % of track distance renewal
- Fastening – fastening clip force mean value
- Sleeper - % of sleeper inspection remarks
- Ballast depth – ballast depth mean value
- Ballast contamination - % of contaminated ballast
- Track quality Q-number
- Track quality c-error - % of track distance
- Track gauge c-limit - % of track distance
- Technical status – weighting of all indicators above according to a redefined analysis model

In a report from Banverket regarding the present state of the governmental owned track infrastructure is several indicators described. The description contains definition, measurement unit, and data source location:

- Number of functional disruptions per track kilometre and year
- Number of functional disruptions per track kilometre and year leading to train disruption
- Number of train delay minutes per track kilometre and year

- Number of train delay minutes per train
- K-number (track quality)
- Q-number (track quality)
- Number of emergency and weekly inspection remarks per track kilometre and year

57 Benchmarking

> **"Benchmarking is the search for best industry practices that will lead to superior performance"**
> (Robert Camp, Xerox Corp.)

Numerous definitions of benchmarking have been developed. Benchmarking is defined as "*a process by which something is measured in relation to best practice*". For Main [98] "*benchmarking is the art of finding out, in a perfectly legal and above board way, how others do something better than you do - so that you can imitate - and improve upon - their techniques*".

Benchmarking is a practical tool for **improving performance** by **learning from best practices** and **understanding the procedures by which they are achieved**. It should lead to a better understanding of costs and is therefore a mean for a better planning and forecasting process and helps to set enhancing but realistic objectives[130].

Benchmarking is a powerful management tool because it overcomes "paradigm blindness"[99]. Benchmarking opens organizations to **new methods, ideas and tools**

[98] How to steal the best ideas around. J. Main. Fortune, 126, (8), p.p. 102-106. 1992.

[99] This phenomenon can also limit our thinking and prevent us from seeing essential factors outside of our paradigm. The pressure to think in a similar way with our fellowmen and women, and pure *paradigm blindness* is lurking us all constantly and everywhere. The claim of the famous physicist lord Kelvin that flying machines, which are heavier than the air,like aircrafts, are impossible, represented a mode of thinking, is an example of *paradigm blindness.*

K. Tzanakakis: *The Railway Track and Its Long Term Behaviour*, STTT 2, pp. 367–368.
DOI: 10.1007/978-3-642-36051-0_57

to improve their effectiveness. It helps **crack through resistance to change** by demonstrating other methods of solving problems than the one currently employed, and demonstrating that they work, because they are being used by others.

If companies are open to changes, competitive benchmarking is most likely to be the most valuable. Radical improvements normally only occur by introducing good ideas from other industries or countries.

Benchmarking is usually used in management and particularly in strategic management; organizations evaluate various aspects of their processes in relation with competitors, usually within their own sector. This allows organizations to develop plans on how to adopt such best practice, usually with the aim of increasing some aspect of performance.

Although benchmarking may be a one-off event, it is best treated as a continuous process in which organizations regularly seek to challenge their practices and follow up their performances over time.

58 Management of Infrastructure Data

For an Infrastructure Managerit is of high importance to have a clear view on the state of its infrastructure.

To be able to manage the maintenance and renewal process efficiently and economically, an Infrastructure Managerhas to set up databases describing the infrastructure and its current state. The databases should provide enough up-to-date information on the infrastructure in such a structured way that the Infrastructure Manager can efficiently use this information to develop the right policy for maintenance and renewal. In case the Infrastructure Manager outsources the M&R projects, the contractors will be responsible for the input of useful infrastructure data. The infrastructure manager's role concerning the data gathering process will become more steering, controlling and facilitating.

In the following, we describe the infrastructure data that should be available in order to make good management possible. It will also elaborate on the data gathering and storage process, first in a situation where the Infrastructure Manageris gathering and storing infrastructure data himself and second in the situation that (a part of) the Maintenance & Renewal (M&R) activities are outsourced[79].

58.1 Infrastructure Data Management of a Rail Infrastructure Manager

An Infrastructure Managershould always keep in mind the following aspects:

- What infrastructure data is available?
- How is it stored (where is it available)?
- How is it obtained?
- How can I use this data to efficiently manage the maintenance and renewal process?

It would be more interesting to ask the question: What information should be available and how should it be stored and analysed?

An abstract from a UIC report[100] provides a useful framework of what kinds of databases are necessary to possess and which kind of data the databases minimally need to consist of.

The following types of databases have been identified:

- The general composition of the infrastructure (system database)
- The history of monitoring operations and the result of these inspections (monitoring database)

[100] UIC "Way &Works" Committee (1996), Maintenance of high speed lines, IF 7/96.

K. Tzanakakis: *The Railway Track and Its Long Term Behaviour*, STTT 2, pp. 369–372.
DOI: 10.1007/978-3-642-36051-0_58

- The history of work carried out (action taken database)
- The history of incidents and faults (incidents and accidents database)
- The inventory of resources and costs (resource and costs database)

The UIC report mentions a list of infrastructure data that minimally should be available in each of the specific databases. The system database should contain a description and decomposition that should be followed in all other databases.

58.1.1 System Database

General characteristics of infrastructure components, for example:

- Curves, connections, cants gradients, transitions
- Structures and earthwork, barriers
- Tunnels, anti-noise walls, cut-and-cover sections
- Singular points (building and various installations, etc.)
- Service and emergency access points
- Drainage systems
- Position of overhead lines
- Switches and crossings locations
- Position of signalling installations

Superstructure inventory:

- Rails
- Ballast
- Fasteners
- Sleepers
- Switches
- Level crossings

To be able to plan maintenance activities for the longterm the information in this database about each infrastructure component should at least include:

- Identification number
- Location (number)
- Building year
- Deterioration models
- Expected lifetime

The identification number should be a unique key ID to be able to store and retrieve information about this specific infrastructure component. This number can be linked to the proposed decomposition of infrastructure and split up according to a certain hierarchy within a component.

The geographical location (as a number) should be included in the unique identification number of the component. This location number can be an important key to link different databases, for example with the use of a GIS interface. In many cases it is necessary to retrieve information on infrastructure components within a certain area, rather than only on a certain type of infrastructure. The location number within the identification number of a component allows for the purpose of linking databases.

For the long-term aspect it is sufficient to know when the infrastructure components have been built and replaced (building year, production year) and what the expected lifetime of the infrastructure component is, according to experiences (deterioration models) and expected use.

58.1.2 Monitoring Database

The monitoring database is important to be able to check, for the long-term, whether the expected remaining service life of the infrastructure component is still feasible. If not, the system should produce a warning to the Infrastructure Managerthat the infrastructure component should be repaired or replaced.

The monitoring database should contain up-to-date information on at least the following subjects:

- Recordings relating to track geometry and ultrasonic inspection
- Work records for switches and crossings
- Evaluation files of the condition of components (sleepers, rails, fastenings, etc.)
- Recordings of corrugation, testing etc.

Two types of data gathering methods feed this database:

1) Data of measurement cars:
Various track condition and geometric parameters are recorded by computerized measurement equipment on board of a special measurement car. Modern measurement cars often include also an ultrasonic inspection of the rails, giving evidence of rail cracks. Other examples are the catenary measurement cars.The measurement cars pass typically every section in the rail network once or twice a year, thus updating the monitoring database.

2) Visual inspections:
Visual inspections also provide input for the monitoring database. Employees of the Infrastructure Manager (IM) walk along the inspection to do visual inspections on the state of the track. If the IM outsources the maintenance& renewal (M&R) activities, these visual inspections are often used to check the work done by the contractors.

58.1.3 Action Taken Database

The action taken database should provide information that enables the Infrastructure Managerto obtain a more realistic view on the remaining service life of a component. From the information in this database the IM should be able to predict for how many years the component should function correctly as a result of the maintenance and renewal (M&R) actions performed (par. 32.6.4).

This database should contain information on the following M&R actions:

- Work done on the geometry of the track and switches (levelling, tamping, aligning, local repairs, lifting, etc.)
- Work carried out on track components (grinding, replacements of sleepers, replacements of rails and switch components, welding of rails, work done on damaged rails, etc.).
- Infrastructure renewals

58.1.4 Incidents and Accidents Database

This database contains an exhausting inventory of incidents, accidents and disturbances as well as their causes and possibly, their remedies.

58.1.5 Resources and Costs Database

This database contains information on all possible resources. Information should be stored on staff (and contractors when the M&R work is contracted out), the equipment, track components, tools, rail and vehicle fleet, and cost management. The last is often stored within an ERP[101] system.

[101] Enterprise resource planning.

59 Maintenance Cost Modeling[102]

59.1 Introduction

When rail vehicles run on a track, the wheels cause tangential and normal forces, as well as vibrations. This results in deterioration, influencing system performance and limiting life of components on both vehicle and track. In order to maintain the performance and safety of the system, the track has to be inspected, maintained or renewed. All these actions result in costs.

A method for estimating costs related to maintenance and renewal of the track is presented in this chapter.

59.2 Maintenance Cost Modelling

Maintenance costs of track infrastructure consist of preventive maintenance cost, renewal cost, and corrective maintenance cost as described in the following. Maintenance costs associated with track have been estimated separately for different curve radii as different curve radii experience different failure probabilities and magnitudes.

Mathematical Formulation

Maintenance costs are the most complex cost component of an asset during its life since maintenance is a long continuous process throughout the asset life. While the cost of any specified maintenance work on an asset can be comfortably estimated using engineering costing methodologies, estimating maintenance costs throughout the asset life is a much more sophisticated process. This is because the types of maintenance are dependent on many factors, of which the most important are asset deterioration rates, maintenance policy, and budget constraint. Maintenance schedules therefore need to be predicted/planned to enable the maintenance costs to be estimated. Maintenance costs of track must include:

- Materials, equipment, and labour
- Condition monitoring and inspection
- Track possession time

In the following, maintenance cost models have been developed with respect to the type of maintenance intervention.

General

M	Life period of track in MGT (Million Gross Tonnes)
m	Gross tonnage per year in MGT
N	Life period of track (equivalent to M) in years

[102] Based on [18].

K. Tzanakakis: *The Railway Track and Its Long Term Behaviour*, STTT 2, pp. 373–376.
DOI: 10.1007/978-3-642-36051-0_59

r	Discount rate
K	Class of curve radii
CL	Average labour cost/hr
L	Total length of track section in km
Li	Length of ith curve in km

Track inspection cost

The cost due to track inspectionprimarily depends on the interval of track inspection and is given by:

$$\sum_{j=1}^{N-1} \frac{((Tt \cdot CL \cdot L) + (Cet \cdot Tt \cdot L)) \cdot \frac{m}{mt}}{(1+r)^j}$$

Tt	Mean time to inspect track in hr/km
Cet	Equipment cost for track inspection /hr
mt	Interval for track inspection in MGT

Track tamping cost

Cost due to track tamping depends on the interval of tamping:

$$\sum_{i=1}^{K} \sum_{j=1}^{N-1} \frac{((Ttai \cdot CL \cdot Li) + (Ceta \cdot Ttai \cdot Li)) \cdot \frac{m}{mtai}}{(1+r)^j}$$

Ttai	Mean time to tamp for ith curve in hr/km
Ceta	Equipment cost for tamping /hr
mtai	Interval for tamping for ith curve in MGT

Rail grinding cost

Cost due to rail grinding primarily depends on the periodicity of grinding and the number of grinding passes and is given by:

$$\sum_{i=1}^{K} \sum_{j=1}^{N-1} \frac{((Tgi \cdot CL \cdot Li \cdot ngi) + (Veg \cdot Tgi \cdot Li \cdot ngi)) \cdot \frac{m}{mgi}}{(1+r)^j}$$

Tgi	Mean time to grind for ith curve in hr/km
ngi	Number of grinding passes on $_i$th curve
Ceg	Equipment cost for grinding /hr
mgi	Interval for grinding for ith curve in MGT

Rail lubrication cost

Cost due to lubrication depends on the number of lubricators in the curves and the cost to maintain each lubricator in terms of filling.

$$\sum_{i=1}^{K}\sum_{j=1}^{N-1}\frac{(Tclu \cdot CL \cdot nli)}{(1+r)^j}$$

Tlu	Mean time to refill lubrication material for each lubricator in hr
Clu	Cost of lubrication material for each lubricator per year
nli	Number of wayside lubricators in *i*th curve

Rail renewal cost

The cost due to rail renewal is given by:

$$\sum_{i=1}^{K}\sum_{j=1}^{N-1}\frac{((Cr \cdot Li) + (Trri \cdot CL \cdot Li) + (Cerr \cdot Trri \cdot Li)) \cdot \frac{m}{mrri}}{(1+r)^j}$$

Cr	Cost of rail /km
Trri	Mean time for rail renewal for *i*th curve in hr/km
Cerr	Equipment cost for rail renewal /hr
mrri	Interval for rail renewal for *i*th curve in MGT

Ballast cleaning cost

Cost due to ballast cleaning primarily depends on the periodicity of ballast cleaning and is given by:

$$\sum_{i=1}^{K}\sum_{j=1}^{N-1}\frac{((Tbi \cdot CL \cdot Li) + (Ceb \cdot Tbi \cdot Li)) \cdot \frac{m}{mbi}}{(1+r)^j}$$

Tbi	Mean time to clean ballast for *i*th curve in hr/km
Ceb	Equipment cost for ballast cleaning /h
mbi	Interval for ballast cleaning for *i*th curve in MGT

Ballast renewal cost

The cost due to ballast renewal is given by:

$$\sum_{i=1}^{K}\sum_{j=1}^{N-1}\frac{((Cb \cdot Li) + (Tbri \cdot CL \cdot Li) + (Cebr \cdot Tbri \cdot Li)) \cdot \frac{m}{mbri}}{(1+r)^j}$$

Cb	Cost of ballasti.e. in €/km
Tbri	Mean time for ballast renewal for *i*th curve in hr/km
Cebr	Equipment cost for ballast renewali.e. in €/hr
mbri	Interval for ballast renewal for *i*th curve in MGT

Sleeper renewal cost

The cost due to sleeper renewal is given by:

$$\sum_{i=1}^{K}\sum_{j=1}^{N-1}\frac{((Cs\cdot Li)+(Tsri\cdot CL\cdot Li)+(Cesr\cdot Tsri\cdot Li))\cdot\frac{m}{msri}}{(1+r)^{j}}$$

Cs	Cost of sleeperi.e. in €/km
Tsri	Mean time for sleeper renewal for *i*th curve in hr/km
Cesr	Equipment cost for sleeper renewali.e. in €/hr
msri	Interval for sleeper renewal for *i*th curve in MGT

Fastener renewal cost

The cost due to fastener renewal is given by:

$$\sum_{i=1}^{K}\sum_{j=1}^{N-1}\frac{((Cf\cdot Li)+(Tfri\cdot CL\cdot Li)+(Cefr\cdot Tfri\cdot Li))\cdot\frac{m}{mfri}}{(1+r)^{j}}$$

Cf	Cost of fasteners in SEK/km
Tfri	Mean time for fastener renewal for *i*th curve in hr/km
Cefr	Equipment cost for fastener renewali.e. in €/hr
mfri	Interval for fastener renewal for *i*th curve in MGT

Track downtime cost

Downtime on the track occurs due to track possession when maintenance actions on the track are executed. Train-free periods (maintenance "windows") are usually used for planning maintenance actions, i.e. the hours between two consecutive trains. However, as the train-free periods are not long enough in most cases, this leads to train cancellations, train speed restrictions etc., which imply penalties imposed on the Infrastructure Managerby the traffic operators. Preventive maintenance and renewal actions are usually planned well ahead so as not to affect the traffic. However, corrective maintenance on the track generally affects the train operation. In our case we have considered rail breaks for corrective maintenance.

$$MTTR\ (rail\ break)=\frac{\sum_i frbi\cdot Trbi}{\sum_i frbi}$$

Trbi	Mean time to repair rail break in *i*th curve in hr
Cer	Equipment cost to repair rail breaks i.e. in €//hr
Mrbi	Mean time to rail breaks in *i*th curve in MGT
frbi	Failure rate of rail (breaks) in the *i*th curve

In this case, track possession time is calculated as the difference between MTTR (see 34.3) and train-free period. Hence, the track downtime cost can be calculated by multiplying the track possession period with the penalty cost.

Part E
The Future

"Study the past if you would define the future"
Confucius

"I never think of the future - it comes soon enough"
Albert Einstein

"Education is our passport to the future, for tomorrow belongs to the people who prepare for it today"
Malcolm X

Section X: Future Challenges for the Infrastructure Manager

Today more trains are running on the track than in the past and the competition with other means of transportation becomes harder. At the same time there is also a tendency towards decreased time and funds for maintenance. The characteristics of the future railway as a system will be discussed in this section.

60 Technological Developments and Tendencies for the Railways

Most of the tracks have been constructed many years before, but the demands put on the tracks today are different from the ones when the tracks were built. During the past decades there were major changes in rail. The maximum axle load increased significantly, as well as the train speed (Figure 179), the loco power and the train density. Today more trains are running on the track and the competition with other means of transportation becomes harder. At the same time there is also a tendency towards decreased time for maintenance and also decreased funds for maintenance.

To face the new situations, more effort has to be put on track maintenance to ensure the issues of safety, comfort, availability and economy.

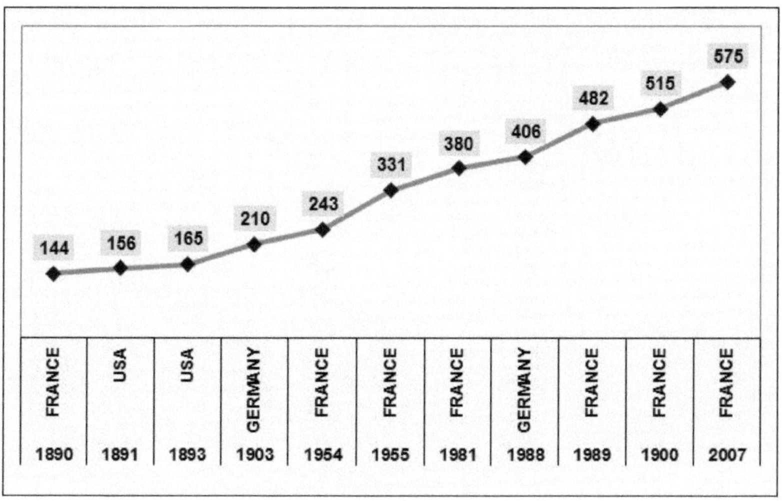

Fig. 179 Speed records

K. Tzanakakis: *The Railway Track and Its Long Term Behaviour*, STTT 2, p. 381.
DOI: 10.1007/978-3-642-36051-0_60 © Springer-Verlag Berlin Heidelberg 2013

61 The Future Railway as a System

The railway is a complex engineering system with long-life assets. Changes require an engineering-led vision of the railway, embodying both (Figure 180):

- changes to the system using the best of current technology and
- a prediction of the changes that may be brought out by new technology.

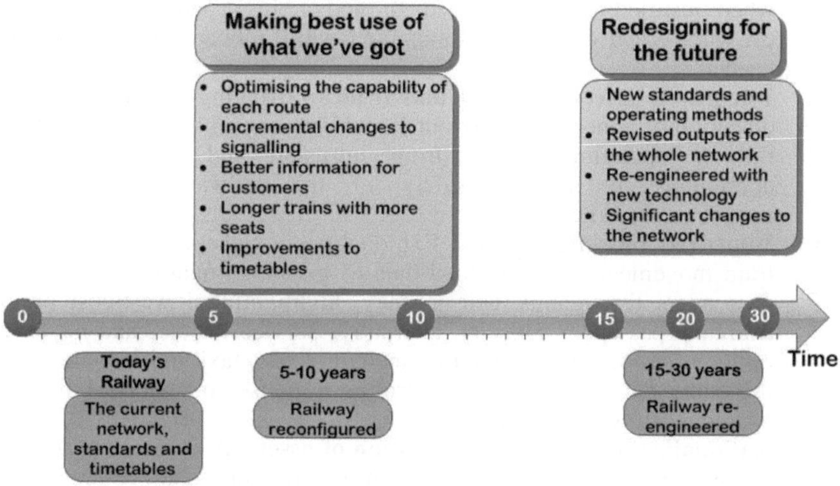

Fig. 180 Developing and delivering the change (after[102])

61.1 Needs and Requirements for the Railway Infrastructure of the Future

The characteristics of the future railway as a system will be [102]:

- **Optimised track–train interface:** Light, but strong rolling stock running on precisely-engineered, accurately-maintained track, reducing energy demand, track maintenance cost and noise;
- **High reliability, high capacity:** Infrastructure designed on lean principles[103] with minimal trackside equipment to minimize scope for critical failures, with monitoring of components that may result in critical failures, so that maintenance is focused where it needs to be and faults

[103] Definition of lean, as developed by the National Institute of Standards and Technology Manufacturing Extension Partnership's Lean Network: "A systematic approach to identifying and eliminating waste through continuous improvement, flowing the product at the pull of the customer in pursuit of perfection."

K. Tzanakakis: *The Railway Track and Its Long Term Behaviour*, STTT 2, pp. 383–391.
DOI: 10.1007/978-3-642-36051-0_61

are detected before they affect service operation, and also promoting availability of the network by minimizing possession requirements;
- **Simple, flexible, precise control system:** Cab signalling combined with an intelligent management layer to reduce equipment costs, combined with an intelligent management layer to offer precise control of train movement through the network, reducing unnecessary stops, facilitating energy efficiency through advisory speeds and allowing its full potential capacity to be realized;
- **Optimized traction power and energy:** Regenerative braking on all trains, whether on the electric network or through hybrids with on-board energy storage. Better use of existing electrification and selective extension where justified by business need. Bi-mode trains capable of running on or off-wire, based on energy storage and with on-board power only where needed;
- **Integrated safety, security and health:** Bringing together automated detection of abnormal behaviour in public areas with better protection of railway operational areas from intrusion and obstruction whilst continuing to improve safety of the railway system through improvements to control;
- **Improved passenger focus:** Exploitation of ticketing, passenger flow, train movement and train load data to give high-quality information to passengers throughout their journey, highlighting lowest-cost options, available passenger capacity, alternative services, routes and modes and options for journey completion, including bus, taxi and bicycle. Use of the same data to optimize controller response to abnormal traffic or passenger flow conditions;
- **Rationalization and standardization of assets:** A rationalized approach to asset specification, with greater use of modular and "commercial off the shelf" (COTS)[104] equipment for industry-specific assets such as rolling stock, based on a whole-life, whole-system cost approach across all industry partners;
- **Differentiated technical principles and standards:** Application of differentiated technical principles and standards to railway routes based on predicted traffic type and usage, allowing cost efficiency to be optimized whilst maintaining interoperability for passenger trains and access for commercial freight to all areas of the network, except where it is agreed that there is no reasonable expectation of future need.

61.2 Filling in the Picture

61.2.1 Optimising the Track–Train Interface

"Steel-wheel-on-steel-rail" technology will remain in use for the foreseeable future, for high-speed and mass public transport as well as for freight.

[104] COTS (commercial off-the-shelf) describes ready-made products that can easily be obtained.

The need to **minimize energy use and track damage, and therefore train mass**, will be a major focus of new train design.

Key changes will be:

- The use of best-practice design to deliver trains with substantially lower mass without compromising safety, combining light body shells, with light bogies, and with interior equipment and furnishings designed within an overall weight target;
- Precisely-engineered, accurately-maintained track, facilitating the development of lighter trains that cause less track damage and reduce maintenance cost, thus creating a virtuous circle;
- Intelligent infrastructure, monitoring its own health and using on-train monitoring so that faults are detected before they affect service operation and help to ensure maintenance is focused where it is needed. The designers of new infrastructure will need to analyse potential failure modes and design-in protection against such failures or, where this is not practical, a process of gradual degradation of service rather than absolute failure;
- Track alignments will need to be maintained to tighter tolerances to achieve the attainable benefits at the track–train interface. The surface damage and wear of rails and wheels can be reduced through a careful balance between track quality, curve radius and cant deficiency on the one hand and wheel profile, conicity and primary yaw stiffness of the vehicle bogie on the other. Sustainable operational limits will be needed for track alignment deviations and wheel wear, driven by system whole-life cost rather than by the much less stringent needs of safety.
- Accurately maintained track to the design alignment will be required to ensure optimum clearances to over-line and platform structures. The need to carry payloads such as 9' 6" containers will necessitate maintenance of adequate clearances to older structures, which are often at a lesser gauge than their modern equivalents. The maintenance of better precision and control of alignment will facilitate the development of sustainable operational limits for the railway, ensuring that best use is made of corridor space and that line speeds are made more consistent.

61.2.2 High Reliability, High Capacity

61.2.2.1 Developing the Present System – "Reconfiguring"

Increasing utilization of existing capacity is a very important theme before increasing the capacity of the system. The systems and incentives need to be put in place to make better use of assets, so that we encourage existing customers to modify their usage of the railway towards quiet, off-peak periods when there are empty seats, empty wagons and even spare train paths available. Peak spreading and better information can help passengers. Better predictability can help freight path take-up. However, these initiatives will always have limits.

Increasing the capacity of the system to carry passengers or freight implies one or more of:

- Increasing passenger capacity or freight payload per unit length of train;
- Increasing the length of trains;
- Increasing the number of trains on the network.

For freight, the major thrust in recent years has been to increase gauge, allowing larger containers and greater bulk to be transported, and this trend will continue. Increasing payload per unit length implies heavier axle-loads, and for bulk commodities such as aggregates on specific routes this is likely to be an economic way of increasing capacity. For most commodities it is likely that lengthening of trains will prove more cost-effective. However, this may require some investment in extended loops, new accommodation and road bridges, and terminals.

Increasing the number of trains on the network may require upgrades at several locations along a route before significant benefits are delivered.

61.2.2.2 Radical Change to the Network – "Re-engineering"

New signaling systems such as ERTMS[105] can improve the potential frequency on individual lines within the network. The ultimate effect will depend upon the characteristics of each route, and particularly what other capacity constraints are encountered. Key junctions, terminal capabilities, and the mix of train speeds all have a role to play, but a "clean sheet of paper" approach to revising the Rules of the Plan will be triggered and should also support the removal of differentiated speeds for freight trains.

61.2.2.3 Network Performance and Availability

Network performance may be improved by reducing the likelihood and impact of primary delay. Reliance on high-reliability equipment will benefit service performance by reduction in likelihood of primary delay.

Work on intelligent, flexible operating rules and efficient operations management may also be of benefit in minimizing the impact as well as supporting provision of additional capacity. This may allow "buffer" capacity to be brought into productive use without increasing overall delay. However, running additional trains in itself presents a greater exposure to secondary delay in the event of any disruption.

High reliability of rolling stock will be essential. Figures of 100.000 miles between service-interrupting failures (between 2 and 4 times the current UK level) are already achieved elsewhere. Reliability will be promoted by an approach based on harmonization of basic elements, standardization of interfaces and evolution of

[105]The European Rail Traffic Management System (ERTMS) is an initiative backed by the European Union to enhance cross-border interoperability and signalling procurement by creating a single Europe-wide standard for train control and command systems.

proven components around a modular concept, such as that being pursued by the European MODTRAIN project[106].

Track maintenance will be planned on a ‘predict and prevent’ basis, based on monitoring assets by train-borne equipment, and use of ‘intelligent infrastructure’ noting trends and rates of deterioration and making planned interventions *before* an asset fails. Access to near-real-time asset status information will allow predictive maintenance to be a reality, greatly reducing network downtime and avoiding premature renewals.

Network Rail is assembling a comprehensive database of information associated with their infrastructure assets and records of its conditions. The data will be available to the asset engineers to assist determinations of deterioration rates and thus maintenance needs. This knowledge bank is populated with the core asset data and data received from track recording trains as well as track-based recording sensors monitoring trains.

61.2.2.4 24/7 Railway

Availability of the network 24 hours a day and 7 days a week (24/7) will be enabled by current and future technology. High-output systems for renewals and ballast-cleaning allow work to be undertaken on a single line with the adjacent line open to traffic under bi-directional working arrangements. Planning around suitable diversionary routes (with equivalent freight capabilities) will be promoted where possible. In the longer term, alternative track formations such as slab track will minimize maintenance needs, especially on critical sections without diversionary routes or where traffic requirements too intensive for single line working have to be accommodated.

New systems, procedures and equipment will be required to provide a 24/7 railways as the norm. This will be achieved through maintenance and renewals being undertaken when the network is more lightly used and reduced capacity would be acceptable. The number of trains, including passenger, empty stock and freight trains, is much reduced in the late evening through to the early morning, a period of 5–8 hours. In this period, the train timetable will be planned to use half the normally available infrastructure, such as two tracks of a four-track section or single line working on a two-track section. On a four-track section, the reduction to two tracks will halve the capacity, subject to the arrangement of lines and their relationship to platforms where the passenger services normally stop. On two-track sections, the capacity will be much less than half and the open line will be working bi-directionally.

Possessions will need to be taken and given up in 10–15 minutes to give time for significant units of work to be completed in an 8-hour period.

---

[106] MODTRAIN stands *for Innovative Modular Vehicle Concepts for an Integrated European Railway System*, and as Integrated Project it is the first of its kind in the area of joint European railway research. The project officially started on 1 February 2004 with a total project duration of four years. MODTRAIN had to define and prove the necessary functional, electrical and mechanical interfaces and validation procedures to deliver the range of interchangeable modules, which will form the basis for the next generation on intercity trains and universal locomotives (http:// www.modtrain.com).

Normal maintenance operations, such as tamping, rail grinding, rail stressing and defect removal, can be undertaken with adjacent lines open. Renewals, however, have always been more problematic. The new high-output track renewals system and ballast cleaners can and do work with adjacent lines open – however, traditional renewals have required a second line for rail cranes. Factory-assembled modular switch and crossing units with pre-commissioned signalling equipment will be installed or replaced within an 8-hour possession.

These systems, work practices and infrastructure designs will enable extension of the maintenance and renewal of infrastructure that can be undertaken with trains running alongside work sites, particularly on two-track sections, and all within 8-hour possessions.

This higher level of availability will be needed despite the additional stress placed on the network through climate change. Higher temperature ranges will require more frequent rail stressing, and more frequent and more intense storms will find weaknesses in drainage and vulnerabilities to embankment slip, bridge scour and wind, particularly for overhead electrified lines.

61.2.2.5 Operational Planning

Traditionally, the railway has relied on knowledge of the network in its incremental approach to the development of timetables. Capacity is determined by experience-based 'Rules of the Plan', which set pragmatic headways and junction margins. When major changes are being carried out, analysis of capacity is made at the scheme design stage, but this is usually related to the changed area only. To maximize the use of latent capacity in the network, a fully integrated approach is needed.

The Infrastructure Managers have to develop a detailed data model of the infrastructure, which will support both engineering change management and operational planning, together with an approach to timetabling based on analysis of train and infrastructure capability using margins set within a 'bottom-up' model. The model will reflect train and infrastructure reliability as well as operational capability. Such a model will also support fast analysis of short-term timetable changes.

61.3 Simple, Flexible, Precise Control System

The current railway control system is complex, has a high infrastructure equipment count, is expensive to maintain and exposes maintenance workers to trackside risks. It is focused on maintaining safe train separation, with little in the way of a system management layer, and is imprecise and hugely variable in terms of the precision of train tracking through the network.

The vision for the future is a system in which intelligent trains each send their own location, derived through data fusion combining satellite positioning data with odometry and balise information, back to a small number of control centres. These will control safety-critical train separation through movement authorities and manage train regulation through driver advisory speeds or timings to minimize delay and save energy.

Track-based train detection will be eliminated, except at key nodes and junctions, where axle-counter detection will facilitate rapid clearance and degraded-mode operations. The (imperfect and opportunistic) ability of the conventional track circuit to detect rail breaks should migrate to infrastructure systems designed for the purpose. Overall, a simple, flexible system will:

- Improve safety, through provision of Automatic Train Protection and reduction in risk to trackside workers;
- Enhance capacity, by decoupling signalling layout from train performance and signal sighting constraints, thus reducing line headways and junction occupation times, with support for Automatic Train Operation where needed for operational reasons;
- Improve service performance, through elimination of primary delay arising from failures in line-side signalling equipment, and through 'gradual degradation' in partial failure conditions;
- Reduce costs, through reduction in the quantity of equipment to be installed and maintained;
- Provide bi-directional signalling as standard to improve flexibility in the event of disruption;
- Facilitate simpler and more optimal electrification schemes by removing signal sighting constraints.

GSM-R will be the operational radio system initially, but this will develop through GPRS to a high-speed operational data-transmission system, working bi-directionally with high availability between train and track.

Future communication-based cab signalling will build on the original ERTMS concept, including both safety and traffic management layers and incorporating satellite-based tracking into the architecture, allowing track-based train detection to be eliminated. Virtually all trains will be fitted with the system unless clearly operating only in yards or on dedicated freight or community lines.

61.4 Optimised Use of Energy

The right long-term solution for rail is one that minimizes its carbon footprint and energy bill.

61.5 An Integrated View of Safety and Security

The track record of the rail mode in terms of safety is good. Despite extensive publicity of accidents, the risk of death or injury from a railway accident is not a major concern for most passengers. Given current trends, personal security in the face of crime and terrorism is likely to be perceived as much more of an issue by passengers in the future and, when passengers decide whether or not to go by train, they see safety and security as aspects of safe travel.

Most major safety risks now arise externally to the rail network – level crossings, vandalism, bridge strikes – and these can be expected to be addressed progressively.

The safety balance will change in the next years:

- Full automatic train protection will be an inherent component of the new control system, which will have been deployed across most of the network;
- Level crossings will be reduced in number and better protected using Closed-circuit television (CCTV) or radar-based obstacle detection, coupled with direct communication with the train, so that train speed can be reduced to a level at which collision with a road vehicle is very unlikely to cause derailment. On the highway side, stricter enforcement coupled with measures such as central dividers at half-barrier crossings should improve compliance;
- Detection of infrastructure failures through measures such as condition-monitoring by fixed equipment or trains carrying measurement equipment will be extensively adopted for economic as well as safety reasons;
- Better redundancy and fall back modes will reduce dependence on the 'fail-safe' principle, which in practice leads to continued operation under manual control and without support of safety systems;
- Error rates and human failures during both train and infrastructure maintenance will be reduced through better testing and better quality control.

Security will be an integral aspect of railway systems, even those not having a safety-critical prime function:

- Access control through continued use of gating on stations will be extended;
- Scanning for passenger numbers and weapons remains an option, although the risk needs to be balanced against the disadvantages in terms of capacity;
- Closed-circuit television (CCTV) will continue to be developed, including automatic face recognition and detection of abnormal behaviour;
- Open train interiors, promoting better visibility of passenger activity;
- External TV and sensors may be employed on trains to detect intrusion.

Concern over public health will be a major issue for public transport, implying demand for:

- Better air conditioning with active filtration of recycled air;
- Anti-bacterial and anti-viral coatings;
- Elimination of uncontrolled-emission toilets.

61.6 Improved Passenger Focus

Although major investment in trains has benefited passengers directly in terms of comfort, cleanliness and reliability, the same level of attention has yet to be paid to passenger information in its widest sense or to stations. Smart ticketing systems are already being rolled out and will rapidly become more integrated with both passenger information and financial systems, initially through the use of mobile phones fitted with Near Field Communications.

Glossary

Term	Chapter	Definition
Asset management	32.4.1	*Systematic and coordinated activities and practices through which an organization optimally and sustainably manages its assets and asset systems, their associated performance, risks and expenditures over their life cycles for the purpose of achieving its organizational strategic plan.*
Availability	34.2	*The ability of a product to be in a state to perform a required function under given conditions at a given instant of time or over a given time interval assuming that the required external resources are provided.*
Benchmarking	57	*A process by which something is measured in relation to best practice.* *The art of finding out, in a perfectly legal and above board way, how others do something better than you do - so that you can imitate - and improve upon - their techniques.*
Capability	34.9	*The ability that the track system will meet its required objectives.*
Corrective maintenance	32.6.3	*The maintenance carried out after recognition and intended to put an item into a state in which it can perform a required function.*
Deterioration or degradation	18	*The reduction of the original quality due to various influences.*
Failure	34.4.1	*Non-conformance to some defined performance criterion.* *The termination of the ability to perform the required function.*

Hazard:	34.2	*A physical situation with a potential for human injury.*
Infrastructure Monitoring	31.2	*The continuous or periodical measurement of the condition of infrastructure components and/or subsystems. Monitoring systems can provide information on the condition or use of the infrastructure. Monitoring information from measurement systems was the main focus of the working group, which devoted less attention to sampling databases or visual inspection data in its studies.*
Life cycle cost	36	*Life cycle cost of an item is the sum of all funds expended in support of the item from its conception and fabrication through its operation to the end of the useful life".* *ISO 15686-5*[107] *defines LCC "the cost of an asset, or its parts throughout its life cycle, while fulfilling the performance requirements".*
Maintainability	34.2	*The probability that a given active maintenance action, for an item under given conditions of use can be carried out within a stated time interval when the maintenance is performed under stated conditions and using stated procedures and resources.*
Maintenance	32.5.1	*Combination of all technical and administrative actions, including supervision actions, intended to retain an item in, or restore it to, a state in which it can perform a required function.* *Maintenance is a set of organised activities that are carried out in order to keep an item in its best operational condition with minimum cost acquired.*

[107] ISO 15686-5 Buildings & Constructed Assets: Service Life Planning: Life Cycle Costing.

Maintenance intervention policies" or "maintenance policies	32.11	*The basic principles by which an Infrastructure Manager is guided in doing maintenance of his assets*
Maintenance plan	32.5.3	*A description of the preventive and predictive maintenance and inspection tasks to be performed at maintenance objects. The maintenance plans describe the dates and scope of the tasks.*
Maintenance planning process	32.5.4	*The process of identifying maintenance needs, contractor capacities, material availability and available budget in order to provide useful and realistic data input for rail infrastructure maintenance planning systems*
Maintenance strategy	32.5.2	*an arrangement of the "types" of maintenance, which is to be done to the track and its components, at a specific rate or frequency, under specific operating conditions*
Operational availability	34.9	*the probability that the track system is operationally available during the train traffic*
Operational reliability	34.9	*probability that during the train traffic operation, the track system will not suffer from any failures*
Organizational strategic plan	32.4.1	*Overall long-term plan for the organization that is derived from, and embodies, its vision, mission, values, business policies, stakeholder requirements, objectives and the management of its risks*

Preventive Maintenance	32.6.2	*A set of activities that are performed on plant equipment, machinery, and systems before the occurrence of a failure in order to protect them and to prevent or eliminate any degradation in their operating conditions.* *The maintenance carried out at predetermined intervals or according to prescribed criteria and intended to reduce the probability of failure or the degradation of the functioning and the effects limited"*
RAM programme	34.2	*A documented set of time scheduled activities, resources and events serving to implement the organisational structure, responsibilities, procedures, activities, capabilities and resources that together ensure that an item will satisfy given RAM requirements relevant to a given contract or project.*
RAMS	34.2	*An acronym meaning a combination of Reliability, Availability, Maintainability and Safety.*
Reliability	34.2	*The probability that an item can perform a required function under given conditions for a given time interval (t1, t2).*
	34.4.2	*Probability that a system or component will work without failure during a specified time interval, with consideration of given operating conditions and environment.*
Renewal	32.5.6	*Any major substitution work on a subsystem or part subsystem which does not change the overall performance of the subsystem.* *The term "renewal" includes all activities involved in replacing a rail infrastructure part or object by a same or similar type of rail infrastructure part or object.*
Safety	34.2	*Freedom from unacceptable risk or harm.*

Service life of an asset	32.10	*Total period during which it remains in use, or ready to be used, in a productive process.* *The assumed period for which a structure or part of it is to be used for its intended purpose with anticipated maintenance but without major repair being necessary.* *A period of time after installation during which a structure (or its parts) meets or exceeds the performance requirements.* *The period of time over which the track provides adequate performance and function with anticipated maintenance but without major repair being necessary*
System	34.1	*Assembly of sub-systems and components, connected together in an organised way, to achieve specified functionality*
System-Effectiveness	34.9	*The probability that a system may successfully meet an overall operational demand within a given time and when operated under specified conditions.* *The ability of a system to do a job for which it was intended.*
Track modulus	16.1	*The stiffness of the spring k per unit length of track.*
Track stiffness	16.1	*The elastic rail deflection that takes place under a wheel loading.* *Track stiffness is the ratio of the load applied to the rail to vertical rail deflection.*
Upgrading	32.5.5	*Any major modification work on a subsystem or part subsystem which improves the overall performance of the subsystem.*

Whole Life Costing	40	*Economic assessment considering all agreed projected significant and relevant cost flows over a period of analysis expressed in monetary value. The projected costs are those needed to achieve defined levels of performance, including reliability, safety and availability.* *The systematic consideration of all relevant costs and revenues associated with the ownership of an asset.*

References

[1] Ferreira, L., Murray, I.N.: Modelling rail track deterioration and maintenance: current practices and future needs. Transport Reviews 17(3), 207–221 (1997)

[2] Jovanovic, S.: Railway Track Quality Assessment and Related Decision Making. Delft University of Technology

[3] Bernhard, L.: Railway Track Optimisation by Efficient Track Maintenance Machinery and Strategies, Research & Testing Department, Plasser & Theurer, Linz, Austria

[4] Larsson, D.: A Study of the Track Degradation Process Related to Changes in Railway Traffic. Licentiate Thesis, Luleå University of Technology

[5] Åhrén, T.: A Study of Maintenance Performance Indicators for the Swedish Railroad System. Licentiate Thesis, Luleå University of Technology

[6] Wenty, R.: Reduktion der Beanspruchung von Gleis- und Zugmaterial

[7] Larsson, D.: Study of the Track Degradation Process Related to Changes in Railway Traffic. Luleå University of Technology (2004)

[8] Holtzendorff, K.: Untersuchung des Setzungsverhaltens von Bahnschotter und der Hohllagenentwicklung auf Schotterfahrbahnen. Dissertation, Technische Universität Berlin (2003)

[9] Martyn, S.D.: Identification and Development of a Model of Railway Track Dynamic Behaviour. Thesis Submitted for the Degree of Master of Engineering, Faculty of Built Environment and Engineering, Queensland University of Technology (2005)

[10] Hawari, H.: Minimising track degradation through managing vehicle-track interaction. Thesis Submitted for the Degree of Doctor of Philosophy, Faculty of Built Environment and Engineering, Queensland University of Technology (2007)

[11] Lichtberger, B.: The Track System and Its Maintenance. RTR Special (2007)

[12] Kohler, M.: Der Bettungsmodul főr den Schotteroberbau von Meterspurbahnen. Abhandlung zur Erlangung des Titels Doktor der Technischen Wissenschaften der Eidgenőssischen Technischen Hochschule Zűrich (2002)

[13] Jovanovic, S.: Modern Railway Infrastructure Asset Management. Delft University of Technology

[14] Track Construction and Maintenance Field Guide, Department of Conservation, New Zealand (2008)

[15] Klotzinger, E.: Der Oberbauschotter Teil 1: Anforderungen und Beanspruchung, ETR NR. 01-02 (2008)

[16] Klotzinger, E.: Der Oberbauschotter Teil 2: Qualitaetsverlauf und Eingriffsschwellen, ETR NR. 03 (2007)

[17] UIC, Maintenance/Renewal Optimisation (MRO)

[18] Patra, A.P.: RAMS and LCC in Rail Track Maintenance. Luleå University of Technology (2007)

[19] IMPROVERAIL, Deliverable 8: Life Cycle Costs for Railway Infrastructure Provision and Use (2003)

[20] Hooymans, M.P., Trompet, M.: Keeping on Track with Maintenance, A holistic view on Rail Infrastructure Maintenance Management (2003)

[21] Krüger, M.: Langzeiteigenschaften der Schotterfahrbahn. EI – Eisenbahningenieur (52) (2001)

[22] United Nations, Economic and Social Commission for Asia and the Pacific. The Restructuring of Railways (2003)

[23] UIC, Asset Management Workshop, Paris (October 27, 2010)

[24] INNOTRACK, Project "Innovative Track Systems"

[24.1] Concluding Technical Report (2010)

[24.2] D1.4.6 - Key to railway infrastructure problems and recommended cost categories (2009)

[24.3] D1.3.3 - Final report on Root Causes of Problem Conditions and Priorities for Innovation (2009)

[24.4] D4.1.5 - Definitive guidelines on the use of different rail grades (2009)

[24.5] D4.2.1 - The impact of vertical train-track interaction on rail and joint degradation (2007)

[24.6] D4.2.4 - Improved model for loading and subsequent deterioration due to squats and corrugation (2009)

[24.7] D4.2.6 GL - Recommendation of, and scientific basis for minimum action rules and maintenance limits

[24.8] D4.5.4 - Friction Management Methods

[24.9] D4.5.5 GL - Guidelines for management of rail grinding

[24.10] D6.4.1- Key values for LCC and RAMS

[24.11] D6.5.4., Guideline for LCC and RAMS Analysis

[25] Fendrich, L.: Handbuch Eisenbahninfrastruktur. Springer (2007)

[26] Esveld, C.: Modern Railway Track. MRT-Productions (2001)

[27] Reddy, V.: Modelling and analysis of rail grinding and lubrication strategies for controlling rolling contact fatigue (RCF) and rail wear. Masters by Research thesis, Queensland University of Technology (2004)

[28] Remennikov, A., Kaewunruen, S., Ikaunieks, K.: Deterioration of dynamic Rail pad characteristics. University of Wollongong, Faculty of Engineering (2006)

[29] Lichtberger, B.: Track Compendium. Eurail Press (2005)

[30] Giannakos, K.: Actions on the Railway Track. Papazissis Publications, Athens (2004)

[31] Kaewunruen, S., Remennikov, A.M.: Investigation of free vibrations of voided concrete sleepers in railway track system. University of Wollongong, Faculty of Engineering (2007)

[32] Iwnicki, S.: Handbook of Railway Vehicle Dynamics. Taylor & Francis Group, LLC (2006)

[33] de Man, A.P.: Pin-pin resonance as a reference in determining ballasted railway track vibration behavior. Delft University of Technology, Faculty of Civil Engineering and Geosdences, Railway Engineering Group, HERON, vol. 45(1) (2000)

[34] Brachin, L., Bruni, S., Lewis, R.: Railway Wheel Wear. Wheel-rail interface handbook. Woodhead Publishing (2009)

[35] Fletcher D., Franklin F., Kapoor A.: Rail Surface Fatigue and Wear. Wheel-rail interface handbook. Woodhead Publishing (2009)

[36] Grassie, S.: Maintenance of the wheel-rail interface. Wheel-rail interface handbook. Woodhead Publishing (2009)

[37] Andersson E.: Models for infrastructure Costs related to Wheel-Rail Interface. Wheel-rail interface handbook. Woodhead Publishing (2009)
[38] Knothe, K.L., Grassie, S.L.: Modelling of railway track and vehicle/track interaction at high frequencies (1993) reprint from NE ISBN 0042-3114; SWETS, Lisse, Z. B.V.: Vehicle System Dynamics. International Journal of Vehicle Mechanics and Mobility 22, 209–262
[39] Chaar, N.: Wheel set Structural Flexibility and Track Flexibility in Vehicle-Track Dynamic Interaction. Doctoral Thesis, Royal Institute of Technology (KTH), Stockholm (2007)
[40] Alias, J.: Characteristics of wave formation in rails. Rail Int. 17(11), 17–23 (1986)
[41] Lewis M.: Railways in the Greek and Roman World. University of Hull
[42] Norman, C., et al.: Design of a system to measure track modulus from a moving railcar. College of Engineering and Technology University of Nebraska – Lincoln
[43] Crawford, S., et al.: Development of a Mechanistic Model for the Determination of Track Modulus, Queensland University of Technology. Queensland Rail, Australia
[44] Selig, E.T.: Track modulus: its meaning and factors influencing it. Transportation Research Record
[45] INNOTRACK, d2111- Methods of track stiffness measurements INNOTRACK Guideline
[46] U.S. Department of Transportation, Federal Railroad Administration. Bridge Approaches and Track Stiffness (2008)
[47] Berggren, E.: Railway Track Stiffness, Dynamic Measurements and Evaluation for Efficient Maintenance. Doctoral Thesis, Royal Institute of Technology (KTH) (2009)
[48] Dahlberg T.: Railway Track Stiffness Variations – Consequences and Countermeasures. International Journal of Civil Engineerng 8(1) (2010)
[49] Banimahd, M., Woodward, P.K.: 3-Dimensional Finite Element Modelling of Railway Transitions. In: Proceedings of 9th International Conference of Railway Engineering, London (2007)
[50] UK, Rail Safety & Standards Board. Review of the effect of track stiffness on track performance (2005)
[51] Reddy, V.: Modelling and Analysis of Rail Grinding& Lubrication Strategies for Controlling Rolling Contact Fatigue (RCF) and Rail Wear. Thesis Submitted for the Degree of Master of Applied Science School of Mechanical Medical and Manufacturing Engineering, Queensland University of Technology (2004)
[52] Lundmark, J.: Rail Grinding and its impact on the wear of Wheels and Rails. Luleå University of Technology (2007)
[53] British RaIlway Board. Permissible Track Forces for Railway Vehicles. Published by Group Standards Railway Technical Centre (1993)
[54] Esveld, C., Kok, A.W.M.: Interaction between moving vehicles and railway track
[55] Carlbom, P.: Carbody and Passengers in Rail Vehicle Dynamics. Doctoral thesis, Railway Technology, Department of Vehicle Engineering, Royal Institute of Technology, Stockholm (2000)
[56] Orvnäs, A.: Methods for Reducing Vertical Carbody Vibrations of a Rail Vehicle. KTH Engineering Sciences, Stockholm (2010)
[57] Fagerlund, J.: Towards Active Car Body Suspension in Railway Vehicles, Department of Signals and Systems, Chalmers University of Technology, Göteborg, Sweden (2009)

[58] Evans, I.: Vehicle Dynamics and the Wheel/Rail Interface, Rail Technology Unit, Manchester Metropolitan University

[59] Iwnicki S.: Simulation of wheel rail contact forces. Manchester Metropolitan University (2003)

[60] UK, Rail Safety& Standards Board. Review of vertical dynamic track forces (2004)

[61] Pyrgidis, C.: Railway Transport Systems, Zitis Editions (available in greek language) (2009)

[62] Okamoto, I.: How bogies work. Japan Railway & Transport Review (December 1998)

[63] Lopez Pita, A., Teixeira, P.F., Robuste, F.: High Speed and Track Deterioration: the Role of Vertical Stiffness of the Track. Proc. Instn. Mech. Engrs., Part F: Journal of Rail and Rapid Transit. 218(F1), 31–40 (2004)

[64] Bona, M.: The effect of straightening and grinding of welds on track roughness. Queensland University of Technology (2006)

[65] Hunt, G.A.: EUROBALT II Final Report to Railtrack: Optimisation of Ballasted Track, AEA Technology Rail report AEATR -TCE-2001-RR-311 (January 2001)

[66] Lopez-Pita, A., Teixeira, P.F., Robuste, F.: High Speed and Track Deterioration: the Role of Vertical Stiffness of the Track. In: Proc. Instn. Mech. Engrs, vol. 218(Part F)

[67] Sharpe, P., Armitage, R.J., Heggie, W.G., Rogers, A.: Innovative Design of Transition Zones. In: Railway Proceedings Engineering (2002)

[68] Hunt, G.A.: EUROBALT IIFinal Report to Railtrack: Optimisation of Ballasted Track, AEA Technology Rail report AEATR -TCE-2001-RR-311 (January 2001)

[69] Gutierrez Romero, M.J., Riley Edwards, J., Barkan, C.P.L., Wilson, B., Mediavilla, J.: Advancements in Fastening System Design for North American Concrete Crossties in Heavy-Haul Service. In: AREMA 2010 Annual Conference & Exposition (2010)

[70] UIC, Leaflet 518, Testing and approval of railway vehicles from the point of view of their dynamic behaviour - Safety - Track fatigue - Ride quality (2009)

[71] Grassie, S.L., Cox, S.J.: The dynamic response of railway track with flexible sleepers to high frequency vertical excitation. Proc. of Inst. Mech. Eng., IMechE 198(D7), 117–123 (1984); see [32], Chapter 6

[72] Read, D., Li, D.: Design of track transitions. Transit Cooperative Research Program. Research Results Digest 79 (2006); Sponsored by the Federal Transit Administration, USA

[73] Murthy D.N.P.: Maintenance of complex systems: issues and challenges. The University of Queensland, Brisbane, Australia

[74] Veit, P.: Track Quality Track Behaviour Track Durability. Graz University of Technology

[75] Veit, P.: Qualität im Gleis – Luxus oder Notwendigkeit? EI – Eisenbahningenieur (57) (December 2006)

[76] Lichtberger, B.: Das System Gleis – Optimierungspotenziale aus Sicht des Instandhalters. Forschungs-& Versuchsabteilung, Plasser &Theurer

[77] IMPROVERAIL, Deliverable 10, Project Handbook (2003)

[78] RailCorp Engineering Manual — Geotechnical, TMC 421 Track Drainage, Version 1.2, (Issued December 2009)

[79] IMPROVERAIL, Deliverable D6: Long Term M&R Policy (2003)

[80] IMPROVERAIL, Deliverable D8: Life Cycle Costs for Railway Infrastructure Provision and Use (2003)

[81] PROMAIN, Life Cycle costs and Maintenance (ProM@in is a Thematic Network of EU, DG Energy and Transport for the enhancement of the performance of Railway infrastructure) (2002)

[82] UIC Leaflet 712, Rail defects (2002)

[83] UIC Leaflet 725, Treatment of Rail Defects (2007)

[84] McMullen, S.R., McCulloch, D.N.: Use of Trench Drains to Improve Chronic Embankment Stability Problems

[85] Australian Rail Track Corporation LTD, DOs and DO NOTs of Subgrade Maintenance

[86] Cantrell Rail Services, Manual on What to Do and What Not to Do When Performing Subgrade Maintenance. Australian Rail Track Corporation LTD (2001)

[87] Selig, E.T., Cantrell, D.D.: Track Substructure Maintenance-From Theory to Practice. American Railway Engineering and Maintenance-of-Way Association Annual Conference Chicago, Illinois, September 9-12 (2001)

[88] Selig, E.T.: Railway Track Design. Handbook of Transportation Engineering (2004)

[89] Kumar, S.: A Study of the Rail Degradation Process to Predict Rail Breaks. Licentiate Thesis, Luleå University of Technology (2006)

[90] Kumar, S.: Reliability Analysis and Cost Modeling of Degrading Systems. Doctoral Thesis, Luleå University of Technology (2008)

[91] Kalousek J., Grassie S.: Rail Corrugation: Causes and Cures. International Railway Journal (July 2000)

[92] Egana, J.I., Vinolas, J., Seco, M.: Investigation of the influence of Rail pad stiffness on rail corrugation on a transit system. Wear 261, 216–224 (2006)

[93] Wenty R.: Hochleistung, Präzision und Zuverlässigkeit in der Gleisinstandhaltung. Eisenbahntechnische Rundschau ETR (2006)

[94] prEN 13848-6, Railway applications - Track - Track geometry quality - Part 6: Characterisation of track geometry quality

[95] Zhang, Y.-J., Alias: FRA Track Quality Indices and Distribution Characteristics. In: AREMA 2004 Annual Conference (2004)

[96] Miura, S., Alias: The Mechanism of Railway Tracks. Japan Railway and Transport Review (March 1998)

[97] UIC, Monitoring Track Condition to improve Asset Management, UIC Asset Management Working Group (2010)

[98] ARTC, Track Condition Monitoring Engineering Process Procedure (PP-146), Australian Rail Track Corporation LTD (2004)

[99] British Standards Institution, PAS 55-1 Asset Management. British Standards Institution (2008)

[100] British Standards Institution, PAS 55-2 Asset Management: Guidelines for Application (2008)

[101] Myburg, J.: Investigating PAS 55: Developing asset management system elements to improve your shutdowns

[102] Department for Transport (UK), Rail Technical Strategy. Published by TSO (The Stationery Office) (2007)

[103] Hooper, R., Alias: Whole-Life Infrastructure Asset Management, Good Practice Guide For Civil Infrastructure. Published by CIRIA, London (2009)

[104] Zarembski, A.M.: Development and Implementation of Integrated Maintenance Planning Systems. Paper presented at the Transportation Research Board Annual Meeting 1998, Washington D.C. (1998)

[105] Veit, P., Wogowitsch, M.: Track Maintenance based on life-cycle cost calculations. ProM@in – Progress in Maintenance and Management of Railway Infrastructure, pp. 6–13 (2002)

[106] Smith, D.: Reliability, Maintainability and Risk - Practical methods for engineers. Elsevier Butterworth-Heinemann (2005)

[107] North South Wales Treasury, Life Cycle Costing Guideline, Sydney (2004)

[108] Barringer, P. (Barringer & Associates, Inc.): A Life Cycle Cost Summary. In: International Conference of Maintenance Societies, May 20-23. Maintenance Engineering Society of Australia (2003)

[109] Blanchard, B.S., Fabrycky, W.J.: Systems Engineering and Analysis, 3rd edn. Prentice-Hall, Upper Saddle River (1998)

[110] Levander, E., Schade, J., Stehn, L.: Life Cycle Costing for Buildings: Theory and Suitability for Addressing Uncertainties about Timber Housing, Department of Civil, Mining and Environmental Engineering, Luleå University of Technology, Luleå, Sweden

[111] van den Breemer, J.J.A.: RAMS and LCC in the design process of infrastructural construction projects: an implementation case. University of Twente, Enschede, The Netherlands (2009)

[112] Patra, A.: Maintenance Decision Support Models for Railway Infrastructure using RAMS& LCC Analyses. Doctoral Thesis Lulea University of Technology Division of Operation and Maintenance Engineering (2009)

[113] Ogink, G., Al-Jibouri, S.: A proposed model for implementing RAMS in the design process in construction (2008)

[114] NASA Reference Publication 1358, System Engineering "Toolbox" for Design-Oriented Engineers (1994)

[115] Todinov, M.T.: Reliability and Risk Models; Setting Reliability Requirements. John Wiley & Sons, Ltd., Cranfield (2005)

[116] Blanchard, B., Fabrycky, W.: Systems Engineering and Analysis. Prentice Hall International, London (1990)

[117] Simoes, G.M.P.: RAMS analysis of railway track infrastructure. University of Lissabon (2008)

[118] EPSMA (European Power Supply Manufacturers Association), Guidelines to Understanding Reliability Prediction (2005)

[119] Rahman, A., Chattopadhyay, G.: Model For Outsourcing Rail Network Asset Using Long-Term Service Contracts

[120] Hirano J.: Outsourcing Transportation Infrastructure Maintenance. MIT

[121] Hauser, J., Katz, G.: Metrics: You are What You Measure! (1998)

[122] Milgate, M.: Alliances, outsourcing, and the lean organization. Quorum Books (2001)

[123] Famurewa, S., Juntti, U., Kumar, U.: Performance-Based Contracting for Maintenance: Towards Achieving Maintenance Objectives. Luleå University of Technology

[124] Campbell, J.D.: Outsourcing in maintenance management: a valid alternative to self-provision. Journal of Quality in Maintenance Engineering 1, 18–24 (1995)

[125] Hyman, W.A.: Performance-based contracting for maintenance: A Synthesis of Highway Practice (2009)

[126] Levery, M.: From Railtrack to Network Rail. IEE Engineering Management (2004)

[127] Jan Swier, J., Slump, T.: Tools and techniques used by ProRail to manage costs and performance, ProRail -The Netherlands

[128] Patra, A., Söderholm, P., Kumar, U.: Uncertainty estimation in railway track life-cycle cost: a case study from Swedish National Rail Administration, Luleå Railway Research Center, Luleå University of Technology, Sweden (2008)

[129] Veit, P., Marschnig, S., Enzi, M.: Preconditions for a Sustainable Low Cost Track. In: 9th World Congress Railway Research (2011)

[130] UIC Project – Lasting Infrastructure Cost Benchmark, LICB (2007)

[131] Åhrén, T., Kumar, U.: Use of Maintenance Performance Indicators: A Case Study at Banverket. In: Proceedings of the Fifth Asia Pacific Industrial Engineering and Management Systems Conference. Luleå University of Technology (2004)

[132] Allander, H.: Objectives Matrix provides information balance. Water Engineering & Management, 144.9, Wilson Applied Science & Technology, 30–31 (1997)

[133] Giannakos, K.: Theoretical calculation of the track-mass in the motion of unsprung masses in relation to track dynamic stiffness and damping. International Journal of Pavement Engineering (IJPE), Special Rail Issue "High-Speed Railway Infrastructure: Recent Developments and Performance" 11(4), 319–330 (2010)

[134] Giannakos, K.: Modern Railway Infrastructure: Resilient Fastenings improve Track's Life-Cycle. In: Conference Transport Research Arena– Europe 2012 (2012)

[135] Giannakos, K.: Heavy Haul Railway Track Maintenance and Use of Resilient versus Stiff Fastenings. In: TRB 90th Annual Meeting proceedings, Washington D.C., U.S.A., January 23-27 (2011); Approved to be published in TRR/2011 (AR060 "Railway Maintenance" Committee)

[136] Fritz, F.: Railroad Track - Theory and Practice. Frederic Ungar Pub. Co., New York (1981)

[137] Eisenmann, J.: Die Schiene als Tragbalken, Eisenbahningenieur (May 2004)

[138] Alias, J.: La voie ferree, IIème edition, Eyrolles, Paris (1984)

[139] Prud' homme, A., Erieau, J.: Les nouvelles traverses en beton de la SNCF, RGCF-2 (1976)

[140] Giannakos K., Loizos A., (2009), Evaluation of actions on concrete sleepers as design loads–Influence of fastenings. International Journal of Pavement Engineering (IJPE) 11(3), 197–213 (November 2009)

[141] EN 50126-1 (Railway applications – The specification and demonstration of Reliability, Availability, Maintainability and Safety (RAMS) – Part 1: Basic requirements and generic process), European Committee for Standardization, CEN (1999)

Index

D

E

S